About the Authors

Rik Drummond heads the Drummond Group, an organization offering electronic messaging consulting to Fortune 1000 companies. The Drummond Group specializes in the implementation of e-mail, EDI, workflow and directory programs, and also provides architectural, record retension, user training, user support, and management procedures.

Nancy Cox has worked for Martin Marietta Information Systems for over eight years in the areas of telecommunications and systems integration. She is the author of "X.500 Directory Services" in *Data Communications for the Office*, and is currently a staff consultant in Advanced Technologies, a research and development group working on large-scale e-mail interconnectivity, directory services, and multimedia integration.

LAN Times E-mail Resource Guide

Rik Drummond
and
Nancy Cox

Osborne **McGraw-Hill**
Berkeley New York St. Louis San Francisco
Auckland Bogotá Hamburg London Madrid
Mexico City Milan Montreal New Delhi Panama City
Paris São Paulo Singapore Sydney
Tokyo Toronto

Osborne **McGraw-Hill**
2600 Tenth Street
Berkeley, California 94710
U.S.A.

LAN Times E-mail Resource Guide

1234567890 DOC 9987654

ISBN 0-07-882052-9

Osborne/McGraw-Hill gratefully acknowledges permission granted by Novell, Inc. to reproduce and include in Chapter 10 excerpts from Novell's *SMF Programmer's Reference Guide*, Copyright ©1992 by Novell, Inc. All rights reserved.

Portions of Chapter 6 are excerpted from IBM marketing materials on IBM OfficeVision/VM (September, 1993) and IBM OfficeVision/400 (September, 1993) by permission from IBM Corporation.

The information in Chapter 12 on Form 1, Variant 1 mnemonic X.400 addressing is provided courtesy of Gary Cannon, a consultant for General Electric Information Services, Inc.

Publisher
   Lawrence Levitsky

Acquisitions Editor
   Jeffrey M. Pepper

Project Editor
   Emily Rader

Copy Editors
   Don Paul
   Polly Russell Kornblith

Proofreaders
   Stefany Otis
   Linda Medoff

Indexer
   Matthew Spence

Computer Designer
   Peter F. Hancik

Quality Control Specialist
   Joe Scuderi

Illustrator
   Lance Ravella

Cover Design
   John Nedwidek

To my best friend and wife, Sara, and my children, Amanda, Rebecca, and Boone

Rik Drummond

To my father, John William Allred (1918-1994)

Nancy Cox

Contents at a Glance

Part I   Introduction to Electronic Mail   **1**

1   An Overview of the Electronic Messaging Environment   3
2   The Structure of an E-mail Application   13
3   E-mail Costs and Benefits   31
4   An Introduction to E-mail Gateways   39

Part II   Getting Electronic Systems to Work Together

5   Connecting to Value Added Networks   53

IIIII 6   Description of Private E-mail Systems 61
IIIII 7   Strategies for Connecting Disparate E-mail Systems . 157

Part III   Electronic Messaging Concepts

IIIII 8   X.500 Directory Systems 175
IIIII 9   IBM SNADS and DIA 187
IIIII 10   Novell NetWare Global Message Handling Service . 193
IIIII 11   SMTP E-mail Services 205
IIIII 12   X.400 Interpersonal Messaging 219
IIIII 13   OSI, LAN, and Other Protocol Stacks 243
IIIII 14   E-mail Network Management 249
IIIII 15   E-mail Gateway Options 255
IIIII 16   Document Conversion Options 267
IIIII 17   What You Should Know About E-mail Security 275
IIIII 18   Electronic Messaging APIs 281
IIIII A   Value Added Network RFP/RFI Technical Skeleton . 293
IIIII   Glossary 305
IIIII   Bibliography 315

Contents

Acknowledgments xxi
Introduction xxiii

PART I

Introduction to Electronic Mail

||||| 1 An Overview of the Electronic
        Messaging Environment 3
        The Generic E-mail Environment 5
        Interconnecting Multiple, Disparate E-mail
            Systems 7
            Services, Features, and Functionality Offered 8
            E-mail Interconnectivity Management 10
        Summary 12

||||| 2 The Structure of an E-mail Application 13
        Overview 14
        The Shared Directory 16

Global Nondistributed Directories 18
Global Distributed Directories 18
Maintaining Global Directory Information 18
Directory User Agent 19
Message Transport Agents 20
The Four Multiplatform Message Transfer
Agents 20
The Message Store/Post Office 21
Message User Agents 22
Gateways 23
The Personal Address Book 24
Access Units 24
Generic Message Format 25
The E-mail Message 26

||||| 3 E-mail Costs and Benefits 31
E-mail Organizational Impact 32
E-mail Costs 33
Forrester LAN Cost Study 33
Ferris Network LAN E-mail Cost Study 33
GartnerGroup E-mail Cost Study 35
GartnerGroup LAN Cost Study 36
GartnerGroup OIS Migration Study 36
Other Studies and Indicators 37
Summary 38

||||| 4 An Introduction to E-mail Gateways 39
E-mail Gateway Basics 40
Types of E-mail Gateways 41
Single Gateways 41
Multisystem Gateways 43
Distributed Multisystem Gateways 44
Common Gateway Features and Services 45
Centralized Directory Facility 45
Directory Synchronization 45
Administration, Configuration, and System
Monitoring 48
Security and Encryption 48
Document Conversion 49
Access to Utilities 49

PART II

Getting Electronic Systems to Work Together

5 Connecting to Value Added Networks 53
Overview 54
Online Services and VANs 55
Gateways to VANs 55
Other Services 57
VANs Offering Both EDI and E-mail Services 57
Vendors 58
Addressing Across the Major Public Systems 58
X.400 Addressing 59
Internet Addressing 59
Proprietary Addressing 59
Criteria for Selecting a VAN 60
Summary 60

6 Description of Private E-mail Systems 61
Banyan Systems, Inc. BeyondMail 63
Product Contact Information 63
Product and Version: BeyondMail
Release 2.0 63
Platforms, Operating Systems, and Network
Specifications 65
Message User Agent 66
Product Directory Structure 66
Addressing Structure 68
Message Transport System 68
Message Structure 69
Envelope Fields 70
Message Store 73
Gateways Supported 73
CE Software QuickMail 74
Product Contact Information 74
Product and Version: QuickMail V2.6 for
Macintosh, DOS, and Windows 75
Platforms, Operating Systems, and Network
Specifications 76
Message User Agent 76
Product Directory Structure 76
Addressing Structure 77
Message Transport System 78
Message Structure 81

Message Store 82
Gateways Supported 82
DaVinci eMail 82
   Product Contact Information 82
   Product and Version: DaVinci eMail V2.5 83
   Platforms, Operating Systems, and Network
      Specifications 85
   Message User Agent 85
   Product Directory Structure 85
   Addressing Structure 85
   Message Transport System 86
   Message Structure 86
   Message Store 86
   Gateways Supported 87
Digital Equipment Corporation's All-IN-1 87
   Product Contact Information 87
   Product and Version: ALL-IN-1 87
   Platforms, Operating Systems, and Network
      Specifications 93
   Message User Agent 94
   Product Directory Structure 95
   Addressing Structure 95
   Message Transport System 95
   Message Structure 97
   Message Store 102
   Gateways Supported 102
Fischer International Emc^2/TAO 103
   Product Contact Information 103
   Product and Version: Emc^2/TAO 3.4 103
   Forms 105
   Platforms, Operating Systems, and Network
      Specifications 105
   Message User Agent 106
   Product Directory Structure 106
   Addressing Structure 107
   Message Transport System 108
   Message Structure 108
   Interface Issues 108
   Gateways Supported 108
HP Open DeskManager 108
   Product Contact Information 109
   Product and Version: HP Open
      DeskManager 109

Platforms, Operating Systems, and Network
    Specifications 111
Message User Agent 111
Product Directory Structure 111
Addressing Structure 115
Message Transport System 116
Message Structure 117
Message Store 118
Gateways Supported 118
IBM OfficeVision/VM 118
Product Contact Information 119
Product and Version: IBM OfficeVision/VM
    V1.2 119
Platforms, Operating Systems, and Network
    Specifications 122
Message User Agent 122
Product Directory Structure 122
Addressing Structure 123
Message Transport System 123
Message Structure 124
Body 125
Message Store 126
Gateways Supported 126
IBM OfficeVision/400 126
Product Contact Information 126
Product and Version: OfficeVision/400 V2
    R3 127
Platforms, Operating Systems, and Network
    Specifications 128
Message User Agent 128
Product Directory Structure 128
Addressing Structure 129
Message Transport System 129
Message Structure 129
Message Store 129
Gateways Supported 129
Lotus cc:Mail 130
Product Contact Information 131
Product and Version: Lotus cc:Mail 131
Platforms, Operating Systems, and Network
    Specifications 133
Message User Agent 134
Product Directory Structure 134
Addressing Structure 135

Message Transport System 138
Message Structure 139
Message Store 140
Gateways Supported 140
Lotus Notes 141
Product Contact Information 141
Product and Version: Lotus Notes Version 3 141
Platforms, Operating Systems, and Network
    Specifications 142
Message User Agent 142
Product Directory Structure 143
Addressing Structure 144
Message Transport System 145
Message Structure 146
Message Store 148
Gateways Supported 148
Microsoft Mail 149
Product Contact Information 149
Product and Version: Microsoft Mail V3.2 149
Platforms, Operating Systems, and Network
    Specifications 151
Message User Agent 151
Product Directory Structure 151
Addressing Structure 152
Message Transport System 153
Message Structure 154
Message Store 155
Interface Issues 156
Gateways Supported 156

||||| 7 Strategies for Connecting Disparate
       E-mail Systems 157
Overview 158
Nonsymmetric Information Mapping 159
Address Mapping Between Systems 159
Information Discrepancies in Inverse
    Mapping 160
X.400 and Internet Message Mapping 160
Heading and Envelope Mappings 160
Body Part Mapping 162
Gateway Functionality Verification 163
Service: Address Mapping 164
Service: Character Set Mapping 164

Service: Limitations on the Number of
    Addresses 165
Service: Subject Field Length 165
Service: Types of Time Stamps 165
Service: Subject Field Contents 165
Service: Message Unique ID/Trace string 166
Service: Priority 166
Service: Content Information—Body
    Part/Content Type 166
Service: Alternate Recipient Allowed 166
Service: Autoforwarded Indication 166
Service: Blind Copy Recipient Handling 167
Service: Body Part Encryption Indication 167
Service: Content Type 167
Service: Conversion Prohibition 167
Service: Conversion Prohibition in Case of
    Loss of Information 167
Service: Converted Indication 168
Service: Cross-referencing Indication 168
Service: Delivery Notification 168
Service: Read Receipt 168
Service: Nondelivery Notification 169
Service: No Nondelivery Notification 169
Service: Return of Content with Nondelivery
    Notification 169
Service: Delivery via Other Distribution
    Means: Fax, Telex, or Postal 169
Service: DL Expansion History Indication 169
Service: DL Expansion Prohibited 170
Service: Express Mail Service (EMS) 170
Service: Multipart Body 170
Problem Resolution 170
Summary 171

PART III

Electronic Messaging Concepts

‖‖‖ 8 X.500 Directory Systems 175
Overview 176
X.500 Directory Components and Protocols 179
Directory Information Tree (DIT) 181
Distinguished Name 182
Access from the DUA 183

Directory Query 183
DSA Cooperative Queries 183
Summary 185

||||| 9 IBM SNADS and DIA 187
IBM's Messaging Architecture 188
Addressing 188
Transport 189
SNADS Message Structure 189
Message Envelope 190
Message Heading 190
Message Body—DIA Structure 191
Address Directory 192
Summary 192

||||| 10 Novell NetWare Global Message
Handling Service 193
Global MHS Message Structure 194
Global MHS Addressing 194
MHS Message Structure 195
Message Envelope 196
Message Heading 198
Message Body 200
Transport 200
Detailed Processing Steps 201
Disk Directory Structure 201
Address Directory 201
Summary 203

||||| 11 SMTP E-mail Services 205
SMTP Messaging Architecture 206
SMTP Addressing 206
Transport 208
Message Structure 211
RFC 822 Message Structure 211
MIME Message Structure 211
Address Directory 217
Summary 217

||||| 12 X.400 Interpersonal Messaging 219
X.400 Messaging Architecture 220
X.400 Addressing 222
X.400 Address Methodology 222

Transport 228
   Message Submission 228
   Message Transfer 229
   Message Delivery 231
IPM Message Structure 232
   ASN.1 Encoding 233
   Envelope 237
   P2 Heading 237
   P2 Body 238
   An ASN.1 IPM Decoded Message 240
Message Store Services 240
Address Directory 241
Summary 241

13 OSI, LAN, and Other Protocol Stacks 243
The OSI Protocol Stack 244
   Application Layer 245
   Presentation Layer 246
   Session Layer 246
   Transport Layer 246
   Network Layer 247
   Data-Link Layer 247
   Physical Layer 248
Summary 248

14 E-mail Network Management 249
E-mail Management Standards 250
   International Organization for
      Standardization (ISO) 251
   International Federation of Information
      Processing (IFIP) 251
   Internet Engineering Task Force (IETF) 251
E-mail Management Framework 251
Management Information Base (MIB) 252
Summary 254

15 E-mail Gateway Options 255
Overview 256
Stand-alone Gateway Solutions 257
Multigateway Switch Solutions 257
Multigateway Distributed Switch Solutions 259
Directory and Transport Backbone Component
   Solutions 260

Product-Independent Message Store Solutions 260
Interpreting the Product Matrix 261
   Interpreting Blank Intersections 263
Summary 263

||||| 16 Document Conversion Options 267
Document Conversion and E-mail 268
   Document Conversion 270
   Conversion Categories 271
Summary 273

||||| 17 What You Should Know About E-mail
      Security 275
E-mail Security Services 276
   Access Control 276
   Message Security Services 277
Summary 279

||||| 18 Electronic Messaging APIs 281
Overview 282
Vendor Independent Messaging (VIM) 282
   VIM Message Structure 283
   Data Attribute Encoding 283
   Character Sets 283
   VIM Simple Message Interface (SMI) 283
   VIM Message Creation Process 284
Common Messaging Calls (CMC) 286
   Sending Messages 287
   Receiving Messages 287
   Looking Up Names 287
   Administration 287
   CMC Messaging Calls 287
   CMC Summary 290
Messaging Application Program Interface
    (MAPI) 290
   MAPI Summary 292
Summary 292

|||||| A Value Added Network RFP/RFI
      Technical Skeleton 293
Technical 294
Delivery Forms Supported 297

Administration 298
Installation 299
Customer Support 299
Costs 300

‖‖ Glossary 305

‖‖ Bibliography 315

‖‖ Index 321

Acknowledgments

Writing a book, especially one covering so many products and topics as this one does, requires the gathering of information with the help of many people from many organizations. We would like to thank the following people for their help:

Ann Wilson, Emily Rader, Mike Bourne, Bill Howell, Bill Moroney, Carol Hamilton, Carol Smykowski, Daniel Blum, David Atlas, David Burch, David Ferris, David Fischer, Debbie Bird, Ed Levinson, Einar Stefferud, Elaine Sharp, Erica Olson, Gary Cannon, Janice Park, Jeff Pepper, Jennifer Goff, Jim O'Gara, Jocelyn Willett, John Finnell, John Fly, John Frederiksen, John Mims, Kit Walsh, Larry Everett, Leslie Schroeder, Mark Elderkin, Mark McHarry, Mary Murphy, Mike McGarr, Paul Bencke, Paul Moniz, Paul Morgan-Witts, Paul Oakland, Peter Heffner, Phil Schacter, Richard D'Alessandro, Richard Madeley, Siobhan Carroll, Tom Burleson, Tom Cavanaugh, Tom Peterson, and Dodi Zelones.

We'd also like to thank WorldTalk and Unified Communications Inc., two electronic messaging switch companies that were especially helpful in reviewing sections of the book.

Introduction

Today's electronic messaging environment is an ever expanding whirlwind of proprietary, pragmatic architectures, as well as a joining of industrial, de facto, and international standards. There are over 50 electronic mail products now being sold and most corporations have more than one electronic mail system installed. Understanding the issues involved in connecting different e-mail systems can be a daunting task. This book will help you to find the information necessary to understand these issues and explain how best to combine multiple e-mail systems into reliable and functional messaging infrastructures.

About This Book

The *LAN Times E-mail Resource Guide* explains the current messaging environment and provides workable strategies for integrating multiple mail systems. The underlying theme of this book is the structure of messages, directories, and message transport systems, in addition to the mechanisms that support each of these structures.

Today, messaging gateways are the key components for implementing highly functional, multivendor, electronic messaging environments. When you select and successfully implement the appropriate gateway, the linked disparate systems work together in a manner transparent to the e-mail system user. Selection of the most technically feasible, cost-efficient gateway depends on the configuration and cost constraints present in your unique messaging environment.

To choose and construct these multivendor networks requires debunking the electronic mail mystique. To help start this process, the *LAN Times E-mail Resource Guide* describes the four basic multivendor e-mail transport architectures in as elementary a manner as possible. The architectures demystified are SNADS, from IBM; SMTP/RFC 822, from the Internet; NetWare Global MHS, from Novell; and X.400, from the International Telecommunications Union (ITU). In addition, this book describes and compares the most commonly used e-mail products, and provides strategies for interconnecting them.

The information in this book is directed at the typical e-mail administrator, messaging network planner, MIS manager, mail-enabled applications developer, e-mail training professional, or end user. This book assumes you have a basic familiarity with an e-mail application and with computer systems in general. If you understand these concepts and architectural components, you have the basic information necessary to construct a highly serviceable, highly functional electronic messaging environment for your company.

How This Book is Organized

To further aid the demystification process, this book is organized into three parts. Part One, "Introduction to Electronic Mail" contains a general introduction to electronic mail, including an overview of the current messaging environment, the structure of a generic e-mail application, the costs and benefits of messaging, and an architectural overview of gateways. Part Two, "Getting Electronic Systems to Work Together" presents strategies for connecting to value added networks (VANs) and implementing gateways, and also provides detailed e-mail system product descriptions. Part Three, "Electronic Messaging Concepts" is a multichapter tutorial on advanced electronic messaging topics such as directories, transport protocols, messaging network management, document conversion, security, and APIs. A brief overview of each of the chapters follows.

Chapter One, "An Overview of the Electronic Messaging Environment," discusses the general nature of electronic mail, defines basic terminology, cites connectivity case examples, and provides assistance on managing e-mail integration projects.

Chapter Two, "The Structure of an E-mail Application," describes the basic structure of an electronic mail application and presents such basic components as directories, transports, user interfaces, message stores/post offices, and the structure of a typical message.

Chapter Three, "E-mail Costs and Benefits," presents the costs and benefits of electronic mail, including organizational impact and migration costs.

Chapter Four, "An Introduction to E-mail Gateways," discusses the architecture of electronic messaging gateways, ranging from simple, point-to-point solutions connecting two different e-mail systems to highly complex multiplatform e-mail switching systems.

Chapter Five, "Connecting to Value Added Networks," provides strategies used when connecting e-mail systems to a value added network (VAN).

Chapter Six, "Description of Private E-mail Systems," contains detailed descriptions of the 12 most common e-mail systems, including DEC's ALL-IN-1, IBM's OfficeVision, Lotus' cc:Mail, Lotus Notes, and Microsoft Mail.

Chapter Seven, "Strategies for Connecting Disparate E-mail Systems," discusses ideas for and challenges in interconnecting e-mail systems, using Internet and X.400 as an example. The chapter also includes examples of information mapping between these two systems and presents a detailed functionality verification test plan.

Chapter Eight, "X.500 Directory Systems," discusses the international standard model for a global, fully distributed electronic messaging directory system, known as an X.500 directory.

Chapters Nine through Twelve discuss the architectures of the four most widely implemented multivendor electronic mail transport protocols: SNADS, Novell Global MHS, SMTP, and X.400.

Chapter Thirteen, "OSI, LAN, and Other Protocol Stacks," discusses in an elementary manner the Open Systems Interconnection (OSI) protocol for data communications between standards-based, open computing systems and relates this model to other protocol stacks such as those used on the Internet.

Chapter Fourteen, "E-mail Network Management," discusses the current state of standards development for e-mail management, including the management framework and the management information base (MIB).

Chapter Fifteen, "E-mail Gateway Options," is a detailed comparison of the market leaders in electronic messaging switches. Products and services from 16 vendors are compared as to their e-mail platforms, transport support, X.500 directory support, and document conversion capabilities.

Chapter Sixteen, "Document Conversion Options," is a description of what to expect when information from one type of computer application program is transmitted via e-mail to an end user with a different program.

Chapter Seventeen, "What You Should Know About E-mail Security," presents current trends in electronic messaging security, such as access control and message security services.

Chapter Eighteen, "Electronic Messaging APIs," is a comparison of the three most prominent electronic messaging application program interfaces (messaging APIs): Vendor Independent Messaging (VIM), Common Messaging Calls (CMC), and Messaging Application Program Interface (MAPI).

Appendix A, "Value Added Network RFP/RFI Technical Skeleton," contains a comprehensive guide for a VAN Request For Proposal/Request For Information (RFP/RFI) to be used when evaluating the service offerings of different public e-mail carriers.

PART ONE

Introduction to Electronic Mail

Chapter One

An Overview of the Electronic Messaging Environment

*C*hris heard a chirping sound indicating that an electronic mail message had just arrived. She leaned over her deck chair to retrieve her palmtop computer, positioning it so the glow of evening light fell on the screen.

"Open new mail," she said and the screen immediately painted a message containing colorful icons. "Open communique," and Chris viewed a full color motion video of Jason, her partner, in their Boston office.

"Hello, Chris!" Jason said. "Sorry to intrude on your much deserved vacation, but I wanted you to be the first to see Alex in action on Baby McGregory's heart surgery. Take a look at this and let me know if we can improve on any of the procedures. The baby's work up and medical records are attached. Have a great time and see you soon."

"Open hologram" and the verbal command created a crisp, full-color, physical representation of a hospital operating room bustling with activity. Leading the team was Alex, the newest member of their cardiac surgical practice. Chris watched, fascinated, while Alex skillfully implanted the latest dynamic growth artificial heart designed to expand over time as the child grew into early adolescence. Then it would be replaced with an adult model.

Chris launched both the baby's chart and medical records to scan them as the surgery progressed. The search capability made it easy to locate any items of specific interest.

"Send reply with video," Chris said, as she looked into the small digital camera in the corner of the screen. "Jason, I have reviewed your information and replay of the McGregory surgery. Please congratulate Alex and his team on an outstanding job! See you on the 7th."

"Send. Close," said Chris, as the image of the operating room evaporated into the cool night air.

Sound far-fetched? Actually, most of the technology used in this message exchange is available today. Although the hologram equipment and the integrated video camera are still beyond the financial grasp of the average end user, it's already possible to send a message containing a variety of information types—such as a spreadsheet, an image, a sound clip, and a video—to a recipient located anywhere in the world. In this chapter, we explore the basic components of the electronic mail, or *e-mail* environment. We discuss the challenge of connecting multiple, disparate electronic mail systems together and identify the management issues involved in any integrated e-mail implementation.

Historically, e-mail systems were used as a productivity tool within a single organization or division to enhance internal communication and distribute large documents electronically. These early systems were proprietary, or vendor-specific, in their architecture and were unable to exchange messages with different messaging systems. However, the increase in the sheer number of e-mail users, the growth of local area networks (LANs), and the frequent mergers and acquisitions in industries all created a need for e-mail systems that could exchange messages with other types of systems in a transparent manner. Over time, organizations have witnessed their e-mail environment shift from an uncomplicated, single mainframe-based system to a highly complex messaging infrastructure composed of multiple, heterogeneous e-mail systems. These sophisticated systems are not only interconnected to each other but also to external e-mail users through public networks.

The Generic E-mail Environment

Electronic mail, or *e-mail*, is the exchange of computerized information in the form of messages. The e-mail application program stores any sent messages in a repository. When a user asks to read his or her mail, the system forwards the messages accordingly. The benefit of this *store-and-forward* message-handling technique is that it uses nonsimultaneous communication. In other words, neither the sender nor the recipient has to be logged onto the e-mail system in real time for the mail to be delivered.

In addition to operating in the store-and-forward mode, all e-mail environments have several basic ingredients. These standard components, illustrated in Figure 1-1, include users, mail messages, sender and recipient addresses, e-mail gateways, protocols, messaging transports, value added networks (VANs), and directory systems. In Figure 1-1, a user on e-mail system A is sending a message. A *user* is an entity that employs the messaging system to send and receive information in the form of structured messages and files. Many users are humans, of course; however, users can also be computer application programs or machine processes that execute certain routines based on instructions contained within messages. Thus, messaging includes user-to-user, user-to-application, and application-to-application information exchange.

Users send and receive messages. A *message* is the actual electronic information transferred between users. A message is structured and usually contains several distinct parts, including a header, a body, and attachments. As illustrated in Figure 1-1, the message has an *envelope* that holds the information necessary to route the message from the source to the destination. The envelope contains the e-mail *addresses*, or location codes, of both the originator and the recipient. An address usually has a proprietary structure that consists of both the user's unique e-mail system identification code and another identifier, such as the e-mail system, the mailbox number, or the organization.

In most cases, each user on the system can access a directory to search for the addresses of a message's intended recipients. An e-mail *directory* usually contains names, addresses, and other demographic data about each user, such as phone numbers, routing information, and preferred delivery method. The international standard for distributed directory systems is the X.500 recommendation, which is discussed in Chapter 8, "X.500 Directory Systems."

After the user creates a message, addresses it, and sends it to the recipient, the e-mail system uses both a message handling protocol and a message transport to deliver the mail to its destination. A *protocol* is a set of rules governing how two like, or peer, entities communicate. An example of a protocol is a conversation between two English-speaking people who both understand what the words, phrases, and letters mean. The same is true in an exchange of messages using the same protocol—the two like entities know what the information looks like, its structure and format, and how to interpret it.

Each e-mail system uses a protocol that describes the structure of the message. The protocol indicates which characters are usable, what the TO: field looks like, and its use. The protocol describes the message structure in at least two parts: the header and the body. The *header* comprises such information as the To:, From:, and Subject: fields whereas the *body* contains the actual message. The body may include text, images (as shown in Figure 1-1), graphics, complex word processing documents, spreadsheets, data, video, audio, and—in the future—items such as holograms. The functionality, services, character sets, and types of reports supported by a particular e-mail system's protocol collectively determine how rich the capabilities may be when you connect different e-mail systems.

The software responsible for moving the message from system to system is called the message *transport*. Examples of transports include X.400, the international standard for messaging, and SNADS, the IBM standard for messaging distribution. Transports are discussed in Chapters 9 through 12. In Figure 1-1, the user on e-mail system A sent the message to two recipients who are on other systems. The message must therefore pass through a gateway to be successfully delivered. A *gateway* is an application program that translates between the different protocols used by two e-mail systems. For example, a software application that enables messages to be exchanged between an SMTP mail system that uses the TCP/IP protocol and a PROFS system that uses the SNADS protocol is a gateway.

A message *switch*, or hub, is a software application that lets more than two e-mail systems exchange messages. Hubs support interconnection to many different message

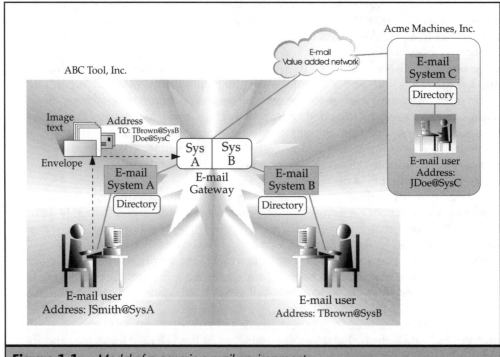

Figure 1-1.   *Model of a generic e-mail environment*

systems and protocols. For example, one message switch may connect cc:Mail, Microsoft Mail, PROFS, ALL-IN-1, and HP DeskManager together in one package.

Value added networks (VANs) are public telecommunications carriers, such as MCI or AT&T, that offer message handling services to subscribers for a fee. VANs enable users in different organizations, such as ABC Tool, Inc. and Acme Machines, Inc., to exchange information. Chapter 5, "Connecting to Value Added Networks," details the connectivity offered by several of these VANs.

Every e-mail system offers the basic functionality described thus far as well as a combination of enhanced services that makes it unique. *Services* are features and functions used by the messaging or gateway system to enable the successful exchange of mail between end users. For example, the ability to reply to a message and to automatically forward it to an alternative recipient are both services. A service description can also include limitations, such as the number of addresses the user can include in a routing or distribution list, to prevent clogging the system.

Interconnecting Multiple, Disparate E-mail Systems

Deciding how to integrate your e-mail systems is a complex task. The integration is specific to your current e-mail environment, resources, and user requirements. Each integration project differs from the next and the products used depend on the particular situation. For example, some organizations have one or two existing e-mail systems to integrate; others have as many as ten systems to interconnect. Some organizations have a clear vision of the next generation of e-mail services to be offered to the users and how these services should be managed. In other organizations, there is little or no input as to what services should be offered or how the different e-mail systems should interconnect—these organizations have no long range strategy.

There are no simple case studies to use as guides when integrating existing e-mail systems. However, there are some general rules that will help you select the appropriate implementation strategy for your particular organization and this book contains the basic information to help you make the necessary decisions. This first chapter discusses these principles and how to use the rest of this book to design the best integration scenario for your environment.

To succeed in both the short and long term, most e-mail integration projects involve three fundamental decisions:

- Which types of services and functionality to offer across the interconnected e-mail systems
- How to manage all the systems to provide the identified services
- Which management techniques to use to surmount the various political and organizational issues normally associated with such a project

Services, Features, and Functionality Offered

The type of services, features, and functionality offered across all internal systems must be identified and "sold" before integration proceeds; otherwise, what you implement may not meet the expectations of the executives and users in the organization. And if you don't fulfill their expectations, the effort will probably be deemed a failure or, at least, a suboptimal project. E-mail integration projects are not necessarily career enhancing—they are political quagmires that require organizational and political expertise in addition to a solid technical foundation. You can simplify the process of determining which services, features, and functionality to provide by answering these questions:

- Will you offer return receipts across all systems?
- Will you have directory services?
- Will you provide document translation?
- Will express delivery be available?
- Will you support the transfer of large files?
- Will you provide delivery and nondelivery reports?
- What level of service are the users anticipating?

All of these questions relate to normal elements of service, which at least some of your users will be expecting after the integration project is complete. The selected services must be documented and clearly conveyed to all concerned so that your implementation matches their needs and expectations.

Two factors determine what services may be offered across all e-mail systems: the e-mail interconnection (gateway) software and the functionality of the existing in-house e-mail systems.

E-mail Gateways

A major ingredient in implementing e-mail connectivity across multiple platforms is the software and hardware used to connect the e-mail systems together. There are over ten message switch, or gateway, vendors in the e-mail connectivity marketplace today. Picking the wrong switch for your implementation may mean you not only pay more than necessary, but that the services you want to support across the organization aren't provided. A message switch is application software that often runs on operating systems such as UNIX, DOS, or OS/2. The message switch software tries to patch the distinct e-mail systems together in as seamless a manner as possible. If done well, the users normally cannot tell that the recipients to which they send messages actually reside on totally different e-mail systems. In such a case, the users don't need to make any special provisions to send mail to recipients on different systems and the addresses look just like their own.

Great switches make the interconnection transparent between all the mail systems. Chapter 4, "An Introduction to E-mail Gateways" introduces the basics of a message switch, while Chapter 15, "E-mail Gateway Options," describes the specific functionality of each one. Appendix A, "Value Added Network RFP/RFI Technical Skeleton," includes sample questions and the requirements typically used in a Request For Information for all major VANs.

Finally, Chapters 8 through 11 serve as a technical tutorial on e-mail architecture components, including the X.500 directory standard and the four message transports: SNADS, Novell's Global Message Handling Service, SMTP, and X.400.

E-mail System Functionality

Each e-mail system supports a slightly different set of services. It's important to select a set of services supported by all systems when they are interconnected. You may choose to implement functions and features common to all systems you plan to connect, such as reply or nondelivery notifications. On the other hand, you may select a functional set from the most robust e-mail system so as not to limit functionality to the lowest common denominator.

For example, suppose several e-mail systems you plan to connect do not support read receipts. *Read receipts* let a sender know that a message was actually read or stored by the recipient. In this case, you might decide not to support read receipts across all connected systems in order to provide a consistent set of features for your end users. (In reality, whether your current set of e-mail systems support read receipts or not, it would be wise to turn off this feature. For read receipts to work well, you must know that they will always work for every user in the organization. Most systems allow the user to specify whether the system should generate read receipts or not. Consequently, you never really know if the read receipt is working or not on a recipient's system.)

For e-mail architects to determine the suggested services for the e-mail integration effort, they must not only understand the message switch architecture but know the services of all the individual systems that will be interconnected.

Detailed descriptions of the most popular e-mail systems are included in Chapter 6, "Description of Private E-mail Systems," to help you determine which architecture and feature set best satisfies your users' requirements. Chapter 6 reviews the basic functionality of these high market-share e-mail systems.

You can examine almost all of the e-mail functionality relevant to an integration project by looking at three aspects of the e-mail system: the transport mechanisms, the directory record structure, and the message structure. These data structures represent the common conceptual thread throughout this book. Whenever possible, the actual fields in the envelope, header, and body of the message are given along with appropriate descriptions. Comparing these fields will give you a solid framework on which to build a robust e-mail infrastructure. It also will help you set the appropriate level of services to offer through all systems.

Adding to the complexity, gateways do not always function as advertised, and currently there is no uniform testing mechanism for gateways. For this reason, we have included a list of the e-mail–specific message conversion functionality scenarios that you should verify before purchasing a switch from any vendor. These scenarios focus on the type of information contained in the messages transferred between the systems, not on the management of the particular message switches or their ease of use. These descriptions, which represent the fundamental tests of message switch conversion functionality, are described in Chapter 7, "Strategies for Connecting Disparate E-mail Systems." (A full set of such scenarios is being developed by the Electronic Messaging Association (EMA) in Arlington, Virginia.)

E-mail Interconnectivity Management

Once you have determined the combination of technical functionality, features, and services that you will offer, you must decide which rules to use to manage the integrated e-mail system so that it fulfills the commitment made to your management and end users. You must apply consistent management procedures and services across all the e-mail systems if you want them to provide a solid, uniform set of services and service levels. Putting these procedures in place is not a trivial task—in fact, it can be harder than managing the technical aspects of the integration in some organizations. To understand why this endeavor may be so difficult, consider the last ten years of the computer revolution. One of the main reasons current e-mail integration is such a challenge is the explosive growth of LAN e-mail systems in corporate America over the last several years. Why has this happened? To put it simply, users were tired of being on mainframes, governed by restrictive MIS policies and procedures with high utilization charges. So they purchased their own personal computers and LANs and began sharing printers, application software, and file servers. This scenario worked well and seemed less costly than the mainframe environment. The user interface was also more colorful, graphical, and easier to use than the historic mainframe interface.

Now, these self-sufficient, workgroup or departmental "islands of automation" are connecting to wide area networks (WANs). Why? Because of the organization's need to use e-mail. E-mail is forcing the "islands of automation" to reconnect so that all users are accessible on the network—but this time they are connecting via corporate-wide networks. However, due to their visibility and cost, users perceive these networks as being like the old, MIS-controlled ones. Highly distributed networks take on the appearance of a "virtual" organization to the end user. Thus, the management rules implemented by your e-mail integration team must balance the need for a uniform service level and set of services (whatever this means in your organization) with the needs of departmental and divisional management. Finding rules that are acceptable to both users and management takes significant expertise.

A word of advice: The implementation of a successfully interconnected e-mail infrastructure requires finesse and great sensitivity. Slapping systems together without agreement and consensus from the users and management involved, even if the CIO or CEO says to do it, will cause ongoing problems in service levels and significant frustration to you and your team. Consensus is the key to making messaging system management work well.

Management Techniques for Successful E-mail System Integration

Project management techniques that help you get through all the political and organizational issues are requirements of any integration project. An integration team should be selected, an executive champion designated, project management software that can track the completion of activities should be procured, and meetings, meetings, meetings should be held to scope and continually redefine the project to align it with changing expectations. Although project management techniques intertwine with the ideas presented here, they require much more time and detail than this book permits. It is sufficient to reinforce the concept that these are high visibility projects that affect many people, are often fraught with politics, and are much more involved than just a technical implementation.

This book contains three major sections whose content becomes increasingly abstract. Part One, "Introduction to Electronic Mail," presents general information about a basic e-mail application, what e-mail gateways are and what they do, and the costs and benefits of e-mail. Part Two, "Getting Electronic Systems to Work Together," contains more specific information on what the major e-mail systems in use today offer as well as what value added networks provide. It also describes strategies for interconnecting disparate systems, including questions to ask the vendors. Part Three, "Electronic Messaging Concepts," presents detailed technical data concerning the basic concepts of the e-mail environment. Keep in mind that it is impossible to maintain an e-mail system without understanding both these basics and the specifics of each product being interconnected.

The key to integrating disparate mail systems is selecting the appropriate gateways or message switches. There are several generic types of gateways: those that connect two systems using a single gateway machine; those that connect several systems using a single gateway machine; and those that connect many systems together by using several machines interacting in a fully distributed manner. Gateways are described in Chapter 15, "E-mail Gateway Options." In addition, use the following scenarios to guide you to the chapters most appropriate for implementing your particular e-mail integration project.

CASE 1:   If your company is only interconnecting a single, well-run e-mail system to one other system or one other VAN, then the issues, while not simple, are less complex than connecting several systems together. If your environment resembles this scenario, and the landscape is unlikely to change, refer to Chapter 7, "Strategies for Connecting Disparate E-mail Systems," to review the primary gateway tests; then see Chapter 15, "E-mail Gateway Options," for a list of vendors and their products. Also, see Appendix A, "Value Added Network RFP/RFI Technical Skeleton," for the appropriate questions to ask a VAN and the functionality tests to perform to verify that the gateway will convey the desired information across the interface. Remember, in today's environment a user *must* verify the gateway's functionality and how it translates data back and forth—what you expect is not always what you get.

CASE 2:   If your company is planning to interface multiple, currently well-run e-mail systems that are geographically and organizationally centralized, then you should consider using one of the various multisystem gateway switches discussed in Chapter 15. Using separate, point-to-point gateways in this environment would incur major management overhead to your company. Multisystem e-mail switches keep the number of components to a minimum by connecting several systems together through a single hardware platform.

CASE 3:   If you will be connecting multiple systems together within a geographically or organizationally dispersed environment, you should consider using one of the large distributed switches described in Chapter 15. Often, you must decide not only which switch vendor to work with but, due to the distributed nature of the deployment, what type of messaging transport protocol to use to interconnect the gateways. Selecting this protocol is tantamount to choosing the messaging transport backbone or infrastructure for your organization. The most prevalent transport protocols are SMTP, X.400, SNADS, and Novell's Global MHS, all of which are discussed in Part Three. The individual systems use their own proprietary protocols. The message switch then converts these proprietary protocols to the interswitch messaging protocol and back again to the target system protocol at the destination. See Chapter 4, "An Introduction to E-mail Gateways," for more details on the architecture of multisystem distributed e-mail switches.

Summary

E-mail has finally reached such a critical mass that users are no longer satisfied communicating only with others on their own e-mail system. These highly competitive times demand that organizations interconnect all their e-mail systems—not only with each other but with users all over the world. This remains a daunting challenge. This book provides the detailed information that system integrators need to accomplish the transparent interconnection of their e-mail environments. Full exploitation of the information age will only be possible when any user can send any information to any other user anywhere at any time.

Chapter Two

The Structure of an E-mail Application

M arket pressure in the LAN environment has forced products to focus primarily on the way the e-mail application presents itself to the end user through the graphical user interface (GUI) and secondarily on internal message transportation and directory synchronization.

Existing international transport or directory standards, such as the X.400 and X.500, and proprietary standards, such as Novell's NetWare Global Message Handling Service (NetWare Global MHS), Simple Mail Transfer Protocol (SMTP), or IBM's Systems Network Architecture Distribution Services (SNADS) have not always been used to implement LAN electronic messaging systems. Over the last several years most new LAN products have implemented proprietary messaging transport and directory systems, often without regard to existing standards.

Adding these new messaging systems to the already existing mini and mainframe e-mail systems resulted in an explosion of different, noncommunicating e-mail products. This chapter defines and discusses the common components of all e-mail systems.

Overview

The conceptual framework on which electronic messaging systems are built is generally the same for all e-mail systems. Components may be missing or combined in some of the architectures, but they keep the same overall structure. The nomenclature of the X.400 and X.500 specifications is used throughout this section. Though not universally followed, these specifications provide a general overall definition of e-mail system requirements and components. Developed by many participants from many organizations, they have garnered worldwide acceptance.

All e-mail systems consist of eight general components. These components provide numerous services to the user, including user address lookup, access control, message security, document conversion, delivery and nondelivery notification, message transportation, message storage, foldering, and message attachments. The eight components, as defined in Table 2-1, can exist on the same computer or on separate computers and communicate across the network. Some e-mail implementations use the distributed client/server model.

Six of the components are shown in Figure 2-1. They are the shared directory, message transport agents (MTAs), the message store/post office (MS/PO), message user agents (MUAs), directory user agents (DUAs), and gateways. The other two, the personal address book and access units, are not a focus of this book and are not shown; however, they are discussed briefly later in the chapter.

Figure 2-1 illustrates the components of an e-mail system based on the X.400 standard as applied to generic mail systems. Callout *a* designates one of two ways an MUA is used to retrieve (or submit) a message from (or to) the e-mail system. In this case, the user's e-mail program attaches directly to the MTA to pickup and submit messages. A special MUA-to-MTA protocol, called Protocol 3 (P3), was established for this purpose in the X.400 standard system.

E-mail Component	Definition
Shared directory	A listing that usually contains the user's full name and e-mail user name (user ID) and other pertinent information such as phone number and message store name.
Directory user agent	A program used to access the directory information that resides on the directory system agent (DSA).
Directory system agent	Generally a distributed database in the e-mail architecture. Holds a partial or complete copy of the total directory information.
Message transport agent	Transfers the mail from the source to the destination addresses.
Message store/post office	Software that supports the user's e-mail filing cabinet. A general purpose server often runs message store or post office software. Some systems use the term "message store" and others use the term "post office."
Message user agent	A program the user employs to interact with the e-mail system to perform various messaging related functions.
Gateway	A program that enables two or more different e-mail systems to exchange messages.
Personal address book	A collection of short user addresses of frequently contacted people.
Access unit	A point of connection between the message transport agent (MTA) and another delivery system such as a fax or telex system.

Table 2-1. *Generalized E-mail System Components*

Callout *b* illustrates the other way in which an MUA may retrieve a message from or submit a message to the transport system. In this case, the message store (MS) interacts with the MTA, on behalf of the user, to send and receive messages. A special protocol, called Protocol 7 (P7) in the X.400 standard, is used between the MUA and the MS. In Internet e-mail systems this type of protocol is called Post Office Protocol 3 (POP3).

Callouts *c* and *d* designate the two routes a DUA may use to gain directory information. Callout *c* shows the process of retrieving information from the MS; callout *d* shows the process of retrieving information from the shared DSA. The protocol used to access the DUA in the X.500 system is the Directory Access Protocol (DAP).

Callout *e* shows interconnections between MTAs. These use a special MTA-to-MTA protocol called Protocol 1 (P1) in X.400, RFC 822 in the Internet system, and ZIP5 in IBM's PROFS (Professional Office System) e-mail system.

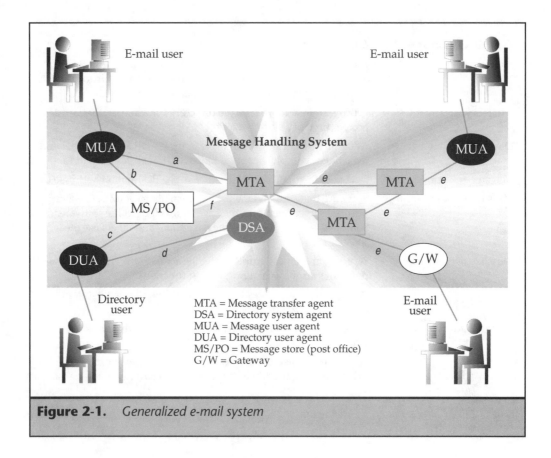

Figure 2-1. *Generalized e-mail system*

Callout *f* designates the interconnection of the MS and the MTA for the transfer of messages in both directions. A special protocol is used between the MS and the MTA. In X.400 this is called Protocol P3.

The Shared Directory

A *shared directory* holds the names, addresses, and other data about the e-mail users and system components. It frequently contains demographic information such as the user's physical addresses, telephone and fax numbers, e-mail address, full name, title, company, and department. Depending on the e-mail product, the message transport system may also use the directory to route mail to the proper destination. This is the case with Novell's NetWare Global MHS product. Also, the directory may contain distribution lists, encryption keys, and document conversion information. Parts of the directory information may exist on each user's user agent, on the MS, or on independent systems that other electronic messaging components use for services.

As a rule, all e-mail systems provide some type of directory facility. The directory is defined and managed by the local system administrator as part of the e-mail system service. It contains a record of everyone on the system and can be viewed and searched by all users. In some cases, this directory supports applications other than just the e-mail application, such as distributed printing facilities, Electronic Data Interchange (EDI), or electronic forms routing.

The X.500 directory was created for globally distributed facilities capable of holding all kinds of information. For example, companies can store phone numbers, user addresses, network IP (Internet Protocol) addresses, advertisements, encryption keys, graphic images of products, and snapshots of people. Frequently the directory resides near the end user to ease the accessing and updating of information. The directory may also be widely distributed to facilitate the search for information about directory items by more geographically dispersed users.

Most directory systems have two parts:

■ Directory information servers, called directory system agents (DSAs), where the information is stored, retrieved, and maintained

■ A program the user employs to retrieve information from the directory, called a directory user agent (DUA)

The result of a typical directory search is shown in Figure 2-2.

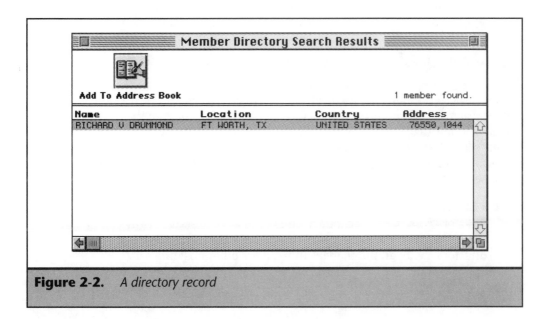

Figure 2-2. *A directory record*

Global Nondistributed Directories

A complete local copy of the shared directory is most often seen in the LAN environment and on older mainframe technologies. For example, cc:Mail and PROFS keep complete copies of the directory on each computer or MS. The local versions of the directory information are updated periodically. When the user requests directory information, the local information is accessed.

Global Distributed Directories

Newer systems have partial local copies of the shared directory. In this form, each MS keeps its directory entries locally, but shares the information with other message stores as required.

An example of this structure is the X.500 directory architecture, which is composed of multiple, distributed, cooperating directory servers called *directory system agents (DSAs)*. This architecture has no single main directory. It does not have copies of the whole directory residing on each system. This architecture is a network of cooperating directory servers, often being maintained by local groups, tied into the overall directory network of tens, hundreds, thousands, or millions of directory components.

The word "local" may be misleading. Some of these local directories, such as the North American Directory Forum, sponsored by most of the North American telephone companies, are local directories for the North American continent, but they are local only in the sense that they do not cover, for example, Asia or Europe.

In this scheme a user community maintains its own directory to support e-mail and other functions, and allows other directories to access information about the local community contained in its directory. Additional information on X.500 directory architecture may be found in Chapter 8, "X.500 Directory Systems."

Maintaining Global Directory Information

The way message stores share local directory information depends on the e-mail product. Microsoft Mail uses a system called Directory Synchronization and cc:Mail uses a system called Automated Directory Update (ADU). Both of these update other servers and message stores with their local subscriber information and automatically update other participating DSAs.

Local user information may be shared with other message stores or post offices by using one of the following techniques:

■ The message store (MS) uses a master directory with multiple secondary directory copies. The secondary directory message stores always update the master MS with their changes. The master MS in turn updates all secondary directories with all the changes from all the other post offices.

- The MS performs updates by sending out mail messages directly to all participating message stores. The updated information is contained in a specially formatted mail message.

- The MS only shares predetermined parts of its local directory information with other directories on other message stores. For example, the local directory may share local registered user names, but not their home phone numbers. This technique is used by X.500 directory systems, which use special protocols to update information.

Directory User Agent

A *directory user agent (DUA)* is the interface between the user and the directory system. The user employs the DUA to view, search, update, and utilize information contained in the directory. A sample DUA is shown in Figure 2-3. In most systems, the DUA is completely integrated with the e-mail package and presents a user friendly and intuitive graphical user interface. Whenever users look up e-mail addresses, they are actually using a DUA to access the directory.

E-mail systems such as cc:Mail, Microsoft Mail, ALL-IN-1, and HP DeskManager are examples of systems in which the DUA is completely integrated with the e-mail software. X.500 systems currently use DUAs that are separate from the message user agents (MUAs). The users may compose a message in their electronic MUA and look up the address by starting the DUA program, to access the directory. In the future, DUAs will be seamlessly integrated into e-mail systems.

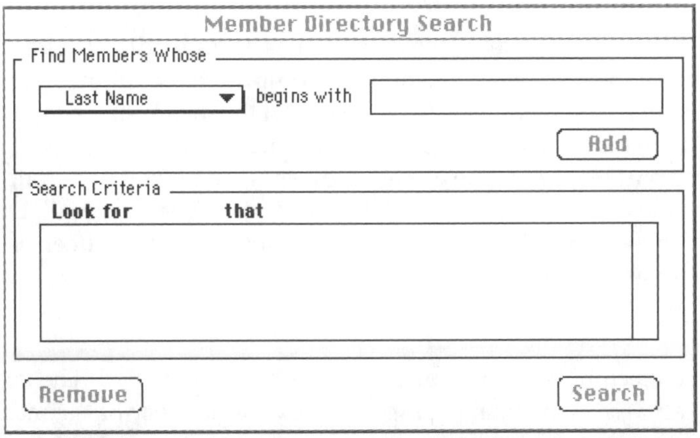

Figure 2-3. *A directory user agent*

Message Transport Agents

The *message transport system (MTS)* controls the way messages are transported and consists of separate *message transport agents (MTAs)*, or software switching units that process and route each message to its appropriate destination. MTAs frequently support distribution list expansion, message splitting, and document conversion.

The message transport is used when a message is transferred over a network or between two message stores. For local messages where the users are on the same computer or share a common MS, the MTA in most products is not invoked to transfer the message. Some other proprietary means is usually employed for local message transfers. Thus, individuals on the same MS receive e-mail immediately after it is sent, while those on other message stores wait for a period of time before the MTA delivers the message. This distinction between how a message is delivered locally versus remotely is generally true across all e-mail architectures.

An example of an MTA would be Microsoft's EXTERNAL program, which executes periodically on a stand-alone DOS personal computer. It checks specified message stores for outbound e-mail. When inter-message store e-mail is found, EXTERNAL copies the e-mail to the destination MS by using file copy routines supported by the underlying network operating system.

The Four Multiplatform Message Transfer Agents

There are four multiplatform MTA standards in existence today. Two of these are international standards: the Internet Simple Message Transfer Protocol (SMTP) and X.400. The other two are proprietary standards widely implemented by several vendors. They are IBM's SNA Distribution Services (SNADS) and Novell's Global MHS. Most e-mail systems use one of these four standards.

SMTP

Internet's SMTP transfers e-mail between message stores by direct session-to-session dialogues between the source and destination computers. When the user selects "send," the message user agent (MUA) program establishes a direct connection to the remote system and transfers the e-mail. The entire message is passed to the destination, one part at a time. The e-mail addresses are passed first, followed by the body of the message. This method is also used by DEC's VMSmail. Due to the size of some of these networks, intermediate transfer points have been established to facilitate the relay of messages.

X.400

This general architecture is the one most frequently implemented. The user hands the message off to an MTA, which takes responsibility for delivery of the message. X.400-compliant systems transfer messages between message stores by passing the messages to the MTA. The MTA then passes the message on through other MTAs, until the last MTA delivers the message to the end-user's system. To facilitate message transfer, the X.400 specification uses an MTA data envelope which surrounds the actual message. X.400 MTAs can expand a distribution list (DL) composed of several recipients of one message, convert a word processing document to a different word processing document format, and by duplicating the envelope as appropriate, deliver it to users on other systems.

SNADS

IBM's SNADS delivers messages by MTAs in much the same manner as in X.400. The SNADS MTAs are called *distribution service units (DSUs)*. The user creates the message and gives it to the DSU to deliver it in the same manner as in X.400-based systems.

NetWare Global MHS

NetWare Global MHS MTAs are called *NetWare Loadable Modules (NLMs)* and they execute on PO servers. These modules transfer e-mail files to remote systems by having a file copy facility supported by NetWare copy them.

The Message Store/Post Office

The *message store (MS)* or *post office (PO)* is a database of created, read, unread, sent, and unsent messages for a single end user. Figure 2-4 shows an example of a message store. The end user sees the MS when logging on to the e-mail system. The MS also supports the in-box and out-box on behalf of the user. In some systems, the user can

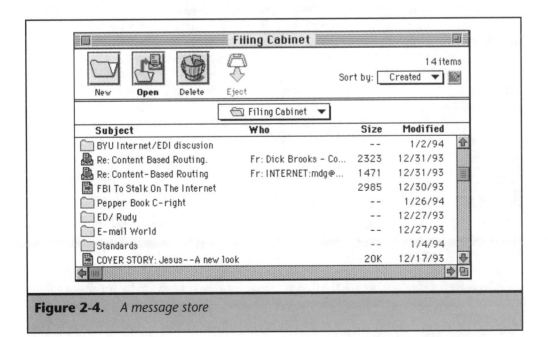

Figure 2-4.   *A message store*

request it to auto-forward, auto-answer, or auto-file messages. Rule-based message sorting and filing, such as that from Beyond Inc., is often implemented on the MS.

An MS serves only one user. Each user can choose to have an MS. The MS acts on behalf of the user and interfaces with the MTA. An X.400 MS is set up so that a user's filing cabinet is physically or logically distinct from those of the other users on the MS. Several vendors have developed interfaces to support the management of the files and messages in the MS.

Inbound messages are deposited in the users' MS and await retrieval. Outbound messages are deposited in the MS by the user. The MS transfers outbound messages directly to other local end users or to the MTA for remote delivery. Normal features supported by the MS (PO) include:

Retrieving messages
Submitting messages
Deleting messages
Filing messages in folders
Listing messages
Auto-forwarding messages based on user-defined criteria
Auto-answering messages based on user-defined criteria
Auto-filing messages based on user-defined criteria
Controlling access by authorized users to messages in the MS

It is hard to tell in some systems where the message user agent (MUA) begins and the MS ends. In older technologies, the programs the subscriber utilizes for e-mail and the MS are often on the same computer. In new products they are separated using the client/server model.

One feature of the MS that clearly differentiates it from the MUA is its availability to the MTA for delivery and redirection of e-mail messages. Even when the user is not logged on or when the user's personal computer is not attached to the network, mail is still deposited.

One of the functions visibly lacking in the X.400 MS is the ability to store messages in user-defined groups or folders, called *P7 foldering*. At this time, the messages just appear in a long list.

Message User Agents

A *message user agent (MUA)* ia a software component that runs on the user's computer and provides the functions of creating e-mail messages, reading received messages, and browsing message lists. Figure 2-5 shows an example of an MUA.

In X.400 and Internet systems, interaction between a remote MUA and the MS or MTA is well defined. The X.400 MUA communicates with the MS by means of protocol P7. On the Internet, the remote MUA communicates with the MS using protocol POP3.

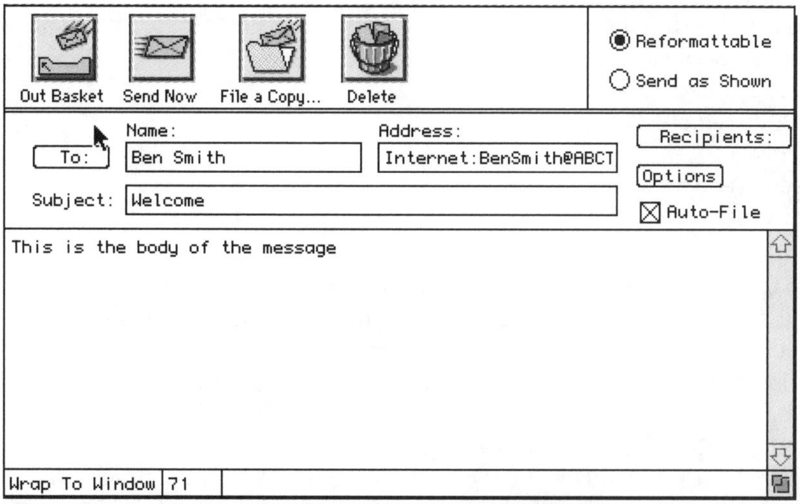

```
┌─────────────────────────────────────────────────────────────────────┐
│   ┌──────┐  ┌──────┐  ┌──────┐  ┌──────┐        ◉ Reformattable       │
│   │  📬  │  │  ✉   │  │  📁  │  │  🗑  │        ○ Send as Shown       │
│   └──────┘  └──────┘  └──────┘  └──────┘                              │
│  Out Basket Send Now  File a Copy...  Delete                         │
│      ▶                                                                │
│            Name:              Address:          ┌─────────────┐       │
│       ┌──────┐ ┌──────────┐   ┌─────────────────│ Recipients: │       │
│       │ To:  │ │Ben Smith │   │Internet:BenSmith@ABCT └─────────┘     │
│       └──────┘ └──────────┘   └─────────────────┐ ┌─────────┐         │
│      Subject: │Welcome                          │ │ Options │         │
│               └─────────────────────────────────┘ └─────────┘         │
│                                                   ☒ Auto-File         │
│   ┌─────────────────────────────────────────────────────────┐ ▲     │
│   │This is the body of the message                          │ │     │
│   │                                                         │ │     │
│   │                                                         │ │     │
│   │                                                         │ │     │
│   │                                                         │ ▼     │
│   │Wrap To Window│71│                                       │ ▦     │
│   └─────────────────────────────────────────────────────────┘       │
└─────────────────────────────────────────────────────────────────────┘
```

Figure 2-5.   *A message user agent*

Many of the MUA-to-MS and MUA-to-MTA protocols are implemented in messaging application program interfaces (APIs) such as MAPI, VIM, and CMC. APIs are discussed in Chapter 18, "Electronic Messaging APIs." These interfaces implement functions to send and receive messages and manage the files in the MS. APIs have components that are in many ways analogous to the X.400 P7 and the Internet POP3 protocols for communicating with message stores.

Gateways

Gateways are application programs that reside on hardware platforms and enable the exchange of messages between disparate e-mail systems. They make it possible for an end user on a LAN-based e-mail system such as cc:Mail to transparently send a message to another end user on a completely different e-mail system, for example, a mainframe system such as IBM's PROFS or a public system such as CompuServe. Originally, gateways linked mainframe e-mail systems, and the connection was fairly straightforward. Today, with the growth of LAN-based systems, multiple gateways for all the various e-mail systems have flourished. The complexity e-mail gateways must handle is staggering. Adding to this complexity, many of the data structures are not well defined or documented in proprietary messaging systems. To maintain any level of e-mail system functionality across multiple systems, e-mail gateways must translate the following types of information between systems:

Character sets
Message heading fields
Message bodies
Message attachments
MTA envelopes
Documents
Special report messages
Addresses

One of the more complex issues pertaining to gateways is the translation of different e-mail system architectures between address formats. For example, the reply-all function does not work well. In some cases, when the reply function is used to respond to a message, the return address is incorrect and the message cannot be routed back to the source. Other gateways do not translate all TO:, CC: (carbon copy), and BCC: (blind carbon copy) addresses.

Chapters 9 to 12 describe the four primary e-mail system architectures. Each is described with respect to address constructs, MTA architecture, message structure, and directory formats. Chapter 6, "Description of Private E-mail Systems," describes the primary e-mail products in terms of these same architectural elements. Chapter 15, "E-mail Gateway Options," describes gateway types and gives detailed functionality comparisons of each of the primary gateway products. These descriptions and comparisons provide the background necessary for you to select the appropriate e-mail gateways for your particular organization's unique e-mail environment.

The Personal Address Book

The *personal address book* is the individual user's private database, and other users are not capable of viewing or searching it. It is not maintained by the directory system and is thus not updated as information changes on the main directory. The personal address book usually contains e-mail aliases that are shorter strings of characters— maybe a first name—instead of the official, and much longer, subscriber addresses. This allows the user to simplify and speed up e-mail address references. An example of a personal address book is shown in Figure 2-6.

Access Units

Access units are points of connection to other mail delivery systems that are not purely electronic, such as fax, telex, teletex, postal delivery, and express delivery. Each access unit usually has some way to accommodate the different addressing needs of the e-mail systems and is predominately outbound from the e-mail system. Recent work has focused on developing international standards to facilitate inbound fax and telex handling by e-mail systems.

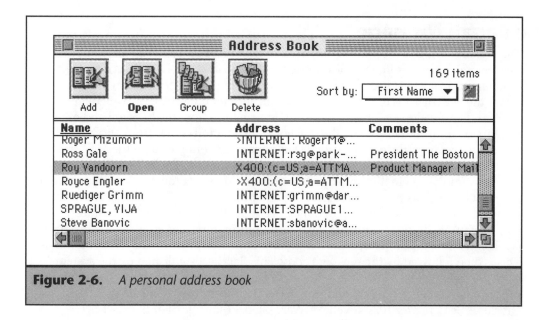

Figure 2-6.   *A personal address book*

As these non-electronic mail systems start supporting both inbound and outbound messages, this distinction will go away, and access units will be called gateways. The term "access unit" is rarely used outside the X.400 standard.

Generic Message Format

In all e-mail systems a message is a message is a message, meaning that all messages must have certain components in order to be created and routed between the originating end user and the receiving end user. All messages have a body, a header, and an envelope. This section focuses on the Interpersonal Messaging (IPM) service, which supports the exchange of generic messages (memos and documents) between end users. IPM services usually contain two types of messages: normal and notification or report messages.

Some IPM services have additional message types, such as the read receipt and the X.400 probe message. The read receipt usually only works between users on the same system. It is not widely implemented in the multivendor environment. The X.400 probe message was designed to verify and test addresses in a nondirectory e-mail environment. Neither of these is common today, nor will they be used much in the future.

The E-mail Message

The e-mail message is normally composed of two major parts: the *heading* and the *body*. Sometimes a third part is also present called an *attachment*, such as a computer file from a spreadsheet program or a graphic drawing. When the message is sent across the network it is often contained within another data structure called an *envelope*. The envelope is constructed from information from the heading and other submission information. Where the envelope information comes from and how it is entered depends on the e-mail system. The envelope is only used by the MTA and only to route the message properly and record information. The MTA usually only changes information in the envelope structure and does not touch other nonenvelope information, such as the message heading and body.

For example, the submission of a message to NetWare Global MHS requires building an ASCII file to establish the header information, and entering the protocol level on the first line, followed by "Sender:," "Addressee:," "CC:," and "Subject:" fields. This is then followed by the message body. This flat file, established by the system software, is transferred into a server disk directory used for outbound e-mail. It is processed and then picked up and transferred to the addresses specified in the envelope. See Chapter 10, "Novell NetWare Global Message Handling Service," for information on NetWare Global MHS.

In the case of an X.400 message, the envelope, header, and body are all distinct data structures within the message. The message heading is built with information such as To:, From:, CC:, BCC:, Date:, and Subject:. See Figure 2-7. The body may be composed of binary or text data. Additional data, over and above the heading and body, is passed to the MTA during message submission. This information may duplicate the heading information: To:, From:, CC:, and BCC:. Additional information such as message priority, deferred submission, and other elements may also be passed. The address information in the heading need not match the information in the envelope. The address fields in the heading can be blank even if those in the envelope contain several addresses. The MTA processes the additional information to build the envelope. The heading and body, while in the same file, are always distinct, identifiable data structures in the X.400 message.

In the case of Internet and NetWare Global MHS messages, the heading ends and the body begins at the first blank line. Other e-mail systems have different methods for identifying the body parts.

Message Heading

The message heading contains fields such as To:, From:, CC:, BCC:, Date:, Reply to:, and Sent:, and describes the type of data contained in the body of the message. An important data element in the message heading is the unique message identification string (or number in some systems). In X.400 systems this is called the *IPM Identifier*. In SMTP, it is called the *Content-Id*. This string uniquely identifies the message from the creator or recipient point of view. It is often used to match messages and responses and to track messages through the various e-mail systems and gateways. Messages of

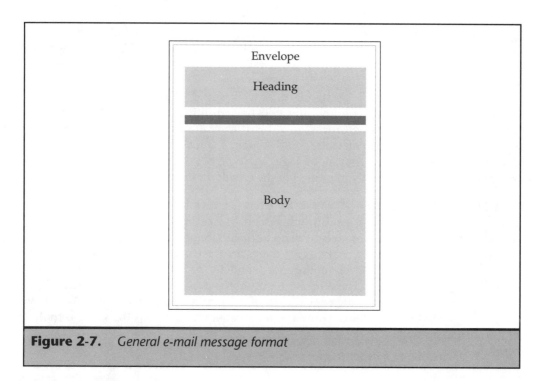

Figure 2-7. *General e-mail message format*

a common conversation, or *threads* as they are called on the Internet, are identified by these message identifiers.

The message heading also contains information required for the end-to-end delivery of messages between e-mail applications. The transport MTA and MS need not understand what is in the heading or body of the message. This distinction can be used to indicate which are heading fields and which are envelope fields. Heading fields are only interpreted by end applications and have no meaning to MTAs. Similarly, envelope fields are used only by MTAs to transfer messages and have no meaning to end applications. MTAs usually only make modifications to the envelope fields. They do not touch the heading fields.

Message Body

The message body historically was composed of ASCII text strings—plain US-ASCII text. Later the ability to attach other messages and binary data was introduced to the general architecture. Many of the LAN systems, such as NetWare Global MHS, still use the attachment method to transport non-ASCII data. The attachment is a separate file from the message, and the user must open the attachment independently of the message. Standard-based system architectures such as X.400, SMTP/MIME (Simple Mail Transfer Protocol/Multipurpose Internet Mail Extension), and SNADS/DIA (Systems Network Architecture Distribution Services/Document Interchange Architecture), include the document within the actual message body; the message and the body part are all one unit. This type of message contains *multiple body parts*. Examples of the attachment method and the multiple body part method are shown in Figure 2-8. See Chapters 9 to 12 for a more detailed description of body parts and attachments in each standard messaging system type.

The 1988 version of the X.400 standard, as well as IBM's DIA and the Internet MIME standard defined during 1992 and 1993 have multiple body part architectures. The architecture allows a single message to carry several different types of information at the same time. The single message could contain a spreadsheet, a word processing document, plain text, and a graphics image. The type of information contained in the separate body parts is usually identified in the envelope structure.

NOTIFICATION MESSAGE   A *notification message* is a status message the sender requests from the MTA when sending a message. Most e-mail systems enable an end user to request status or automatically send notification messages to the sender as a matter of course. Three types of status messages are available depending on the particular e-mail product: delivery notification, nondelivery notification, and forwarded to another user notification. These messages often have a heading and body that are different from normal e-mail messages. They often require special handling when they are exchanged between systems. However, their general structure is the same as that of the normal message.

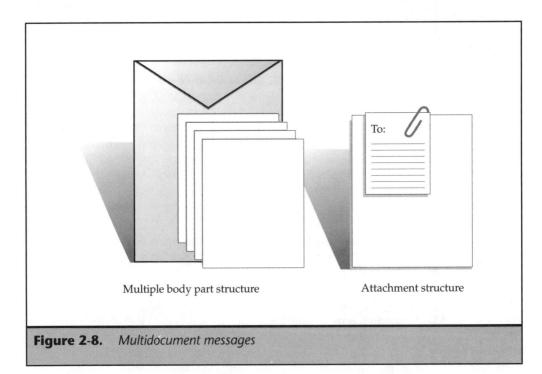

Multiple body part structure   Attachment structure

Figure 2-8.   *Multidocument messages*

The Message Envelope

The message envelope is often composed of information such as: To:, From:, Date:, Subject:, information about delivery report requests, return route for the message, and who to reply to if not the sender. See Chapters 9 to 12 for a description of e-mail envelope formats used in SNADS, NetWare Global MHS, SMTP, and X.400, respectively.

The envelope is used only by the MTA to move messages through the system and determine the type of services the user is requesting. The MTA modifies the envelope as necessary to include the return path of the message, expanded addresses through a distribution list, and forwarding requests. The envelope is the portion of the document that specifies the type of services the originator of the message is requesting from the mail system. In most systems the envelope is the only part of the entire message the MTA modifies. The envelope is implemented differently depending on what transport system is being used: X.400, SMTP, SNADS, or NetWare Global MHS.

In some of these systems, the information in the envelope surrounds the message header and body; in others the envelope information is mingled with that of the message heading. An example of this mingling is shown in the SMTP/MIME message in Figure 2-9. The structures of those with distinct heading and envelope fields, IBM's SNADS and X.400, are not so intuitive. The structure of their messages is described in Chapters 9 and 12, respectively.

```
Sender: ifip-emailmgt-request@ics.uci.edu
Received: from ics.uci.edu by odin.xma.com id af27850; 11 Jan 94 15:52 PST
Received: from ics.uci.edu by q2.ics.uci.edu id aa16880; 11 Jan 94 15:11 PST
Received: from ietf.cnri.reston.va.us by q2.ics.uci.edu id aa16855;
Message-Id: <199401112051.AA25678@zephyr.isi.edu>
To: IETF-Announce:;@compuserve.com
Subject: RFC1566 on Mail Monitoring MIB
Cc: xxx@isi.edu
Mime-Version: 1.0
Content-Type: Multipart/Mixed; Boundary=NextPart
Date: Tue, 11 Jan 94 12:51:19 PST
Sender: ietf-announce-request@ietf.cnri.reston.va.us
From: "Joyce K. Reynolds" <jkrey@isi.edu>


—NextPart


A new Request for Comments is now available in online RFC libraries.
This ..............


—NextPart—
```

Figure 2-9.   *An SMTP/MIME message*

In Figure 2-9 the Received:, Message-Id:, Mime-Version:, and Content-Type: fields are envelope fields. The To:, Subject:, Cc:, Date:, and From: fields are heading and envelope fields.

Review

E-mail messages are normally composed of three parts:

- The *envelope,* which the MTA uses to transport the message and record the MTA user services requests
- The *heading,* which contains address information and dates and times that are used by the e-mail applications to intercommunicate
- The *body,* which may have several general forms

E-mail gateways are complex and require sophisticated programming to handle all three parts of the message structure, addresses, and MTAs. Gateways are thoroughly described in Chapter 4, "An Introduction to E-mail Gateways."

Chapter Three

E-mail Costs and Benefits

Until recently many business executives were unsure if e-mail was a wise or even a necessary business investment. The costs and specific benefits were hard to pinpoint, because e-mail was usually just one of several applications on an existing computer platform. The objections to e-mail have receded as companies realize that if managed and implemented properly, e-mail can facilitate business decision making and enhance productivity. The larger and more geographically dispersed the company and the more time zones involved, the more e-mail has shown a return in easing the communications burden.

Five years ago most e-mail was on mainframes such as HP Desk Manager, DEC ALL-IN-1, IBM PROFS, and Emc^2/TAO. Today, e-mail on LANs surpasses e-mail on mainframes in the spread of new installed mailboxes. The most recent market survey by the Electronic Messaging Association (EMA) estimates there are over 8.9 million mailboxes in the Fortune 2000 companies alone and that most of the growth is taking place in the small branch sites rather than at headquarters.

MIS managers estimate that e-mail on the mainframe costs between $500 and $1,000 a year per user. The study found that because many of them were already using LAN systems anyway, they decided to go to LAN-based e-mail systems, rather than continue using links to mainframe systems. This seemed, initially, to be a less expensive route than mainframes. Subsequent findings have shown that ill-planned and ill-managed LAN environments can cost several times more than mainframe e-mail systems. This is due to the higher integration and maintenance costs of LANs. Studies do not always agree on the actual costs, because they use different starting points and assumptions, but their results show that LAN e-mail systems are at least as costly as mainframe e-mail systems.

E-mail Organizational Impact

E-mail flattens the communications structure of organizations by allowing correspondence in a less hierarchical manner. This helps avoid the communication bottlenecks that plague traditional decision-making management structures. E-mail saves time by reducing the number of phone calls and helping with scheduling meetings. It can also make people more effective in their day-to-day activities.

E-mail is a tool that when used appropriately:

■ Enhances the breadth of communications across organizations

■ Facilitates communications across time zones

■ Speeds and improves overall communications

The argument about whether e-mail is necessary is no longer an issue. The question now is, "what are the real costs of e-mail?"

E-mail Costs

Several recent studies have tried to identify and compare costs on LAN versus mainframe support systems and measure up the costs of running an e-mail application on either one. Such studies have often been conducted as part of downsizing efforts so prevalent in large United States companies. Three premiere research and consulting firms, Forrester, Ferris Networks, and the GartnerGroup, provide the best comparative analyses.

Forrester LAN Cost Study

A Forrester study on overall LAN costs, not just e-mail, which was reported in the January 4, 1993 issue of *Network World*, found that the average 5,000-user LAN costs $6.4 million per year to support, or $1,280 per user. On average, $750 (59 percent) of this is used by the LAN administrator doing routine support tasks.

Table 3-1 illustrates the costs of supporting a LAN and all its applications, not just e-mail costs. In addition, these figures are just for support and do not include depreciation and maintenance on end-user devices, infrastructure hardware, and software. Dave Ferris of Ferris Networks believes that about 30 percent of LAN support costs may be attributed to e-mail. In this study, that works out to $384 per user.

Ferris Network LAN E-mail Cost Study

In 1993, Ferris Network detailed in the *Ferris Electronic Mail Analyzer* newsletter, the cost of a LAN and the incremental e-mail applications and support costs. The study estimated the cost to support a LAN workstation to be $2,500 a year per user. Table 3-2 shows how the LAN network platform costs were computed.

Ferris Network then examined the incremental cost of adding e-mail to the LAN platform. The cost for a small 50-user site was barely ten percent higher per mailbox than the 1,000-user site. The *Ferris Electronic Mail Analyzer* showed that the cost for an

Support Type	Number of Support Staff	Support Person per Number of Users	Support Cost per User
LAN administration	67	1 per 75 users	$750
Help desk	8	1 per 625 users	$100
Physical LAN support	19	1 per 260 users	$280
Other			$140

Table 3-1. *Forrester LAN Cost Study*

Platform Installation Costs	50 Workstations	1,000 Workstations	10,000 Workstations
Network design and planning	$4,400	$66,000	$330,000
Hardware and software products	$150,000	$3,000,000	$30,000,000
Cable installation	$15,000	$300,000	$3,000,000
Technical support classes	$3,000	$37,000	$375,000
Total platform installation costs	$172,400	$3,403,500	$33,705,000

Platform Annual Operating Costs	50 Workstations	1,000 Workstations	10,000 Workstations
Technical support	$25,000	$500,000	$5,000,000
Technical support classes	$3,000	$37,500	$375,000
Additional products	$5,000	$100,000	$1,000,000
Product maintenance	$32,500	$650,000	$6,500,000
Communications	$2,500	$50,000	$500,000
Total annual platform operating costs	$68,000	$1,337,500	$13,375,000

Total Platform Costs	50 Workstations	1,000 Workstations	10,000 Workstations
Three-year platform costs	$376,400	$7,416,000	$73,830,000
Platform costs per workstation per year	$2,509	$2,472	$2,461

Table 3-2.   *Ferris Network LAN E-mail Cost Study*

e-mail account per year is $389 for a small site, $349 for a large, single site (due to volume site licensing), and $448 for a large, multisite installation. These figures exclude the cost of the underlying PC and network. Table 3-3 shows how the incremental e-mail costs were computed.

E-mail Installation Costs	1000 Mailboxes	6000 Mailboxes
E-mail software products	$70,000	$600,000
Message store	$160,000	$1,600,000
Message router on a PC	$4,000	$124,000
Dial-in PC gateway	$4,000	$28,000
Fax gateway	$39,000	$285,000
Public e-mail gateway	$600	$1,800
PROFS	$4,000	$4,000
SMTP		$2,000
Total e-mail installation costs	$281,600	$2,644,800
Annual Operating Costs	**1000 Mailboxes**	**6000 Mailboxes**
Local PC network support	$170,000	$1,020,000
Central technical support	$20,000	$120,000
Communications	$25,000	$300,000
Public e-mail services	$3,500	$21,000
Product maintenance	$36,608	$343,824
Total annual e-mail operating costs	$255,108	$1,804,824
Total E-mail Costs		
Three-year e-mail costs	$1,046,924	$8,059,272
E-mail costs per mailbox per year	$349	$448

Table 3-3. *Incremental PC E-mail Costs*

Ferris Network believes that about $350 per year of the LAN platform costs should be allocated to the e-mail application. This would bring the cost of a mailbox to between $700 and $800 per year.

GartnerGroup E-mail Cost Study

A GartnerGroup report, "Electronic Mail Cost of Ownership: A Chargeback Perspective" by M. Anderson (*OIS Research Notes,* August 9, 1993), discusses the amount charged back to an e-mail user for use of the system. Anderson found that the amount charged back often is not the total cost of e-mail, because e-mail use is frequently subsidized in some manner by the organization.

The GartnerGroup estimates that the average yearly cost per user on a Windows workstation attached to a LAN is $13,500. The cost of e-mail support is about $4,600 a year per user. The GartnerGroup concurs with Ferris Network that about 30 percent of the traffic on a LAN is attributable to the e-mail application, and feels that LAN-based e-mail costs are at least 30 percent more than mainframe-based e-mail costs. These costs could be considerably higher, depending on the hardware, software, and communications required for the transition from a mainframe to a LAN system.

GartnerGroup LAN Cost Study

A GartnerGroup report titled "Buy a LAN and Pay for Three" by C. Wegmann (*OIS Research Notes*, August 9, 1993), examines the tremendous costs involved in setting up LANs. The annual network cost per end user in a tightly-coupled corporate LAN varies between $5,500 and $7,250. These costs do not take into consideration the end-user terminals and hardware and software costs. The GartnerGroup considers the major costs to be the following:

Category	Type	Percent of Total Cost
Components	Cabling, network adapter cards, hubs, bridges, routers, network operating systems, file servers, and installation	36
Connectivity administration	Planning, support, and maintenance of the physical infrastructure	18
NOS/applications support	Network servers, systems management, software distribution, network software, applications license management, and some end-user technical support	22
Hidden costs	Changing user interfaces, end-user training, workstation upgrades, and new system operations	24

GartnerGroup OIS Migration Study

Another GartnerGroup report, "OIS Migration: Cost Analysis Results" by M. Anderson (*OIS Research Notes*, August, 9, 1993), suggests that moving e-mail to LANs to reduce costs makes sense only if it is well planned. The analysis compares LAN-based and mainframe-based e-mail systems.

Anderson found that the cost of providing e-mail to a mainframe user is $970 per year, while for a LAN it is $1,300 per year. The main reason LANs costs are higher is that it takes over two and a half times the cost to maintain system integrity in the LAN environment as it does in the mainframe environment.

Other Studies and Indicators

Several years ago, Drummond and Associates conducted a five-year cost of ownership study on e-mail platforms in a campus environment of 2,500 users. The study focused on the annual equipment infrastructure, software installation, and maintenance costs for all components on two mainframe systems. These costs included terminals but not user or system support. The results were as follows:

Vendor/Product	User Device	Average User E-mail Cost per Year
DEC/ALL-IN-1	Terminal	$504
IBM/PROFS	Terminal	$684

In another example, the 20,000-user ALL-IN-1 network of a major defense contractor ran at an average of less than $500 a year per user. That cost included everything from floor space, hardware, software, hotline, training, network, and part of the moves, adds, and changes.

An INDEX group study, "The Cost of Network Ownership," by Dr. Michael E. Treacy compared a defense contractor's network costs to the costs of 11 other networks and found that the defense contractor's costs were about 40 percent below the others. The average cost of e-mail in that study came to about $900 a year per user in a normal mainframe environment.

One of the reasons LAN-based e-mail and other applications often cost more than mainframe-based applications may be indicated by the lack of importance information managers place on infrastructure. *Network World* surveyed 160 network LAN managers to determine what they looked for in LAN e-mail applications. Table 3-4 shows the rank of importance the managers gave to e-mail features. Even though the managers ranked standards as the third most important item, they had the infrastructure features of directory synchronization, gateways, X.400 support, and fax integration ranked way below the user-visible areas of application integration. The reason LAN-based e-mail costs so much more than mainframe-based e-mail is exactly because of the lack of attention to detail paid to the architecture, design, management, and operation of LAN e-mail systems. LANs seem simple, and they are simple compared to mainframes, but as the number of devices increases, the LAN's operational complexity increases exponentially, surprising users, administrators, managers, and executives. If information managers applied as much effort to the construction of the LAN environment as they did to the mainframe environment, LAN and client-server costs would become more competitive.

Most LAN and LAN e-mail vendors are working to increase operating efficiency and alleviate LAN management costs by implementing standards in the

E-mail Feature	Rating
Security	4.47
Application integration	4.21
Adherence to standards	4.16
Price	3.83
Reputation of vendor	3.83
Directory synchronization	3.70
Availability of gateways	3.69
Fax integration	3.09
X.400 support	3.07
Video, voice, and other	2.54

Table 3-4.   *E-mail Feature Ratings by LAN Managers*

infrastructure. Both Microsoft's EMS e-mail and the Lotus LCA-compliant architecture for Notes and cc:Mail utilize programs to implement the X.400 and X.500 suite of protocols. These are positive signs and indicate solutions that in the near future will partially alleviate exorbitant LAN management costs.

Summary

E-mail costs for LAN-based systems can be much higher than e-mail costs for mainframe-based systems. This is attributable to the extra costs of operating a LAN as compared to a mainframe. Vendors are working to reduce these costs by implementing standard-based message transport agent (MTA), directory system agent (DSA), and message store (MS) components. The cost of LAN e-mail will drop as these infrastructure components become available in the near future.

Chapter Four

An Introduction to E-mail Gateways

As e-mail systems have proliferated, the need to exchange messages between disparate systems in a transparent and reliable manner has become paramount. In the past, mainframe e-mail systems connected easily with a software program running on one of the mainframes. This capability, generally between two well-established e-mail systems such as DEC's VMSMail and IBM's PROFS, bridged the two systems and enabled the exchange of basic text messages. These one-dimensional messaging bridges, known as *single*, or *point-to-point*, gateways have become somewhat obsolete with the advent of LAN-based e-mail systems. As businesses expanded or LANs replaced existing mainframe systems, LAN e-mail users required connectivity with increasing numbers of e-mail systems. It was not uncommon in the late 1980s for large U.S. corporations to have eight or more different e-mail systems functioning and operating at once in different divisions or departments. However, at that time any connectivity generally occurred only between the company's mainframe systems.

Once companies achieved basic connectivity, users expected their e-mail systems to be capable of sending the computer files they generated locally (word processing documents, spreadsheets, graphics, and so on) to other users on totally different e-mail systems. What's more, the users wanted the files to be received unaltered—"what you see is what THEY get"—and in a timely manner. Mainframe e-mail bridges could not accommodate these requirements, so e-mail gateways were developed to go beyond the exchange of simple text messages to include compound documents with audio and video capabilities. Gateways represent the key technology in the successful deployment of a reliable and accurate messaging transport service within a heterogeneous e-mail environment.

This chapter provides a high-level orientation to e-mail gateways, including alternative architectures for the deployment of both point-to-point and multiple gateways in increasingly more complex, growth-oriented settings. Also discussed are how gateways function within an e-mail environment, the role of directory synchronization performed at the gateway level, and the typical features gateways offer.

For a detailed comparison of specific products from major gateway vendors, refer to Chapter 15, "E-mail Gateway Options."

E-mail Gateway Basics

E-mail gateways are mainframe- or LAN-based application programs designed to facilitate the exchange of messages between two or more different e-mail systems. Gateways translate addresses, message headings, and body parts between the systems. When a message is sent, the gateway processes the message from the originating or source e-mail system, changes the formats of the source system into those required for the recipient, or *destination*, system, and then transfers the message to the destination system. In this way, cc:Mail users, for example, can send messages as well as attached files to Microsoft Mail users just as easily as they can send messages to users within their same cc:Mail post office.

The gateway examines both the sender and recipient e-mail addresses in order to route the message and provide a path on which the recipient's reply, delivery, or nondelivery notification can travel back to the sender.

The gateway usually contains some level of directory information that may include listings of message store/post office names and subscriber addresses for both e-mail systems. When these names and addresses are changed, added, or deleted from the end-system directories, the gateway may perform a directory synchronization between the two e-mail products. This way updates on one system are automatically reflected in the gateway's directory, and messages can continue to flow between the two systems.

Types of E-mail Gateways

E-mail gateways can be categorized into three main groups based on their respective complex functions: single, multisystem, and distributed multisystem.

Single Gateways

Single, or point-to-point, gateways connect just two e-mail systems and are obviously the least complex and easiest gateways to implement and manage. They can run on either a personal computer or mainframe platform. This type of gateway is suitable for small organizations with a relatively homogeneous e-mail system base and little growth in disparate LAN-based systems. For example, a company with a population of PROFS users that wishes to connect to the Internet would benefit from a point-to-point gateway. The cost of a single gateway ranges from a few thousand dollars for a PC version to $50,000 for a mainframe version.

Figure 4-1 is an abstract view of a single gateway e-mail system interconnection between VMSmail and PROFS, two mainframe-based systems. As shown, a message goes through several stages before being delivered to a user on another e-mail system. First, the user on VMSmail creates an e-mail message and addresses it to a subscriber on another e-mail system (stage 1)—in this case, PROFS. The message is transferred to the single gateway, called "Exchange1," over a data communications line (stage 2). Exchange1, the single gateway, picks up the incoming e-mail message (stage 3), performs the required translations (stage 4), and routes it over another communications facility (stage 5). PROFS, the destination e-mail system, places the incoming message in the intended recipient's in-box (stage 6).

A very elemental text message requiring the use of a gateway is depicted in Figure 4-2. The e-mail message is originated by Dick Smith, a Microsoft Mail user, and addressed to two other users: Sally Jones, who is on the same e-mail system, and Jeff Pepper, who is within the same company, but is a subscriber on an SMTP system. This is a simple message, with no graphic or spreadsheet attachments.

When Dick sends the message, Microsoft Mail copies it to the outbound, shared mail directory. In this case, it makes two copies; one for Sally and one for Jeff, because they are on different systems. Sally's copy of the message is sent to her local file server,

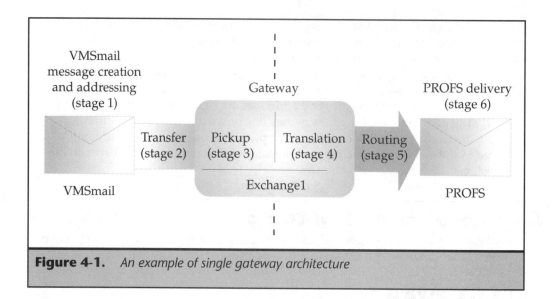

Figure 4-1.   *An example of single gateway architecture*

and Jeff's message is copied to a single gateway such as Exchange1, shown in Figure 4-1. While the message exchange process is now complete for Sally, the local recipient, the message destined for Jeff has a little more work to do.

Jeff's message now resides on the PC running the single gateway software, which is both an e-mail server and an Internet TCP/IP host. The gateway translates the message and gives it to an SMTP program to make the final delivery to Jeff. The gateway performs message translations from Microsoft Mail format to SMTP format, converting dates, addresses, subject field, body field, and character sets. The addresses of all three users, the sender and both recipients, must be translated to SMTP format so that when Jeff replies to the sender, Dick, a copy will also go to Sally.

```
Date:***
From: Dick Smith Dick_Smith@ WorkGroup1.ABCTool
To: Sally Jones Sally_Jones@ WorkGroup1.ABCTool
To: Jeff Pepper Jeff_Pepper@ WorkGroup2 ABCTool.co:
Subject: Test Message

This is a test. This is a test.
```

Figure 4-2.   *An example e-mail message*

The SMTP program locates the name of the host containing Jeff's e-mail box by looking in Jeff's local system's address tables. If the addresses match, the message is transferred one address at a time, followed by the rest of the message, to the remote system. After the transfer is complete, the link is closed and the local SMTP program delivers the message to the end user. A more detailed example of an SMTP exchange is given in Chapter 11, "SMTP E-mail Services."

Multisystem Gateways

Multisystem gateways connect more than two e-mail systems and work much like single, point-to-point gateways except that they have multiple inputs and outputs. The Retix OpenServer 400, as shown in Figure 4-3, is an example of a multisystem gateway. In this sample network configuration, the OpenServer 400 manages the connectivity between a cc:Mail and an SMTP e-mail system. The OpenServer can connect a variety of e-mail environments, including, for example, cc:Mail, Microsoft Mail, SMTP, and QuickMail. Multisystem gateways, also called *e-mail switches*, usually have centralized features such as a central directory facility, synchronization among all the participating e-mail systems, security and access control, and document conversion facilities.

Hub-like gateways become cost effective as the number of e-mail systems requiring connectivity increases. At three e-mail systems, putting in a multisystem

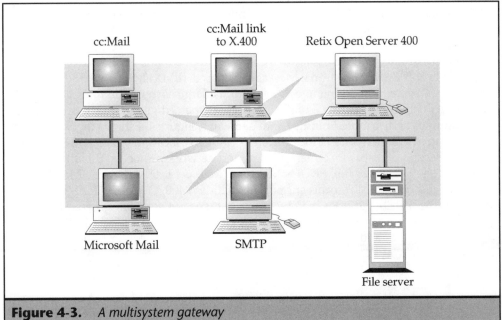

Figure 4-3. *A multisystem gateway*

gateway is normally appropriate, due to the need for centralized directory services, management, and document translations.

The messages are processed in much the same manner as shown in Figure 4-1, except the hub now examines the destination address of the message to decide which new formats, if any, to use in converting the message's envelope, heading, and body parts. The gateway may have to make several copies of the message to send to the different e-mail systems. It may also have to convert documents from one format to another, such as convert a WordPerfect document to a Microsoft Word document. (Single systems can also convert documents, but only if a conversion utility is supplied.)

Gateways functioning as hubs have "plug and play" modules for each e-mail system and supported protocol. This means that the modules can be connected to the hub without requiring further programming. The multisystem gateway must also support the required physical and network connections for each e-mail system connected. For example, it could implement SNA, OSI, and LAN Manager links at the same time.

Distributed Multisystem Gateways

Distributed multisystem gateways work the same way as the gateway described above, except that multiple hubs are located throughout the organization's e-mail network. Distributed multisystem gateways function in either a master/slave or peer-to-peer configuration. The gateways work together to maintain the central directory, coordinate directory synchronization across the network, facilitate the administration and monitoring of the system, and provide end-to-end reliable message transport to all network locations.

These hubs have a two-tier decision tree, unlike a single hub, which receives a message from one e-mail system and hands it off to the other system. A distributed hub decides whether to route the message locally among its own participating e-mail environments or transfer the message to the next appropriate hub on the network.

Multiple Soft*Switch Electronic Message Exchange (EMX) hubs and DEC's MAILbus 400 systems, working in concert within a large corporation to facilitate connectivity between disparate e-mail systems in dispersed geographic locations, are good examples of distributed multisystem gateways. Figure 4-4 shows the Soft*Switch EMX system distributed among four different company locations, each with its own unique collection of disparate e-mail systems and network protocols. E-mail systems shown include SMTP, cc:Mail, PROFS, Microsoft Mail, ALL-IN-1, Banyan Mail, VMSMail, and HPDesk. Protocols supported include X.400, DECNET, TCP/IP, and OSI. All of the distributed hubs may be centrally managed and configured. This approach is also used by companies using e-mail in mission-critical applications to provide redundancy and alternate routing. Such a distributed configuration could cost as much as $250,000.

Common Gateway Features and Services

Most gateways currently on the market offer a common set of features and services. Gateway vendors differentiate themselves by the number of different e-mail systems their product can connect, the ease of installation and administration of their product, their adherence to international standards such as X.400 and X.500, and by offering unique enhancements and features, such as automatic directory synchronization or systemwide monitoring.

Centralized Directory Facility

All gateways allow a participating e-mail system to map its proprietary directory to the gateway's central directory facility. This enables messages to flow from one e-mail system to the other, based on unique addresses contained in the central directory.

Users are generally provided with a query-by-mail facility that enables them to search for addresses by sending an e-mail query to the central directory. The user may specify a recipient's last name, department number, or location, for example, as search criteria. The directory then responds by sending a message back containing all matching entries in the directory. The entries contain the information necessary for the user to identify the intended recipient and to correctly address the e-mail message. Usually, either the central directory has one record per end user containing all the user's known e-mail addresses, or each different e-mail system contains such a record for each user.

Some gateways permit mail to flow through them regardless of whether the sender is registered in the central directory. This facilitates the easy exchange of e-mail. Other gateways require that all users, senders and recipients, be registered in the central directory. This requires considerable administration but gives the organization a measure of control over who is exchanging e-mail with whom. Gateways can offer automatic registration or a registration utility to aid in filling in the central directory.

Directory Synchronization

A major feature of gateways is *directory synchronization,* the ability to synchronize the directories of two different e-mail systems. This is no small task since most e-mail system directories are proprietary and their naming conventions vary considerably. The most advanced gateways adhere to the X.500 directory standard, which uses features such as common fields, attributes, field lengths, and a standardized manner in which to perform directory queries, updates, and reads. Each gateway hub on a network using the X.500 standard maintains its own local directory and routes all queries through the network to the appropriate hub for resolution.

Figure 4-4. *Distributed multisystem gateways using Soft*Switch EMX*

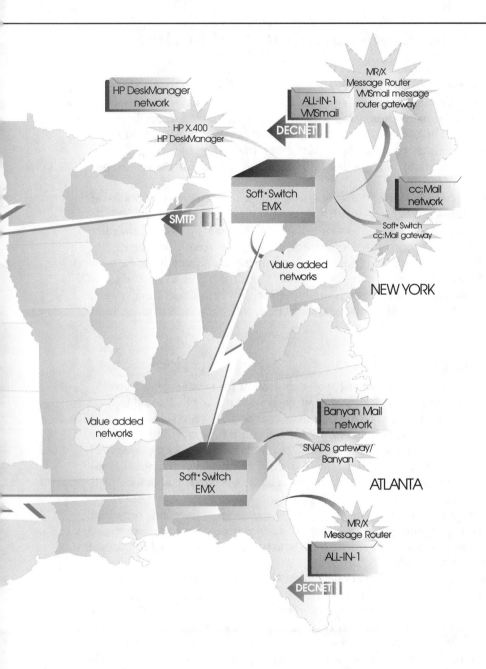

Administration, Configuration, and System Monitoring

Most gateways provide administrative tools that enable system managers to effectively and efficiently configure new e-mail systems, administer the network from a central location, perform automatic directory synchronization, and perform system back up in the event of a failure. They also offer programs that enable a system administrator to monitor the flow of e-mail through the system by checking the number of messages in the mail queues and assess how quickly messages can be delivered to each of the end systems.

Gateways also provide utilities that enable an administrator to perform routine system maintenance, troubleshooting, and system diagnostics. Some systems offer remote diagnostic capabilities that allow the vendor to diagnose the gateway through an online dial-up facility. Also, the user interface for these capabilities has recently become more intuitive and graphical, shielding the administrator from the underpinnings of the network or software program. Today, most gateway administrators do not need to know the UNIX programming language to manage an e-mail network. To perform routine system configuration, monitoring, and maintenance, they need only know how to navigate through the administrative screens.

Security and Encryption

Gateways offer security in the form of access control for the systems that connect to them and for the end users of the gateway. Each system must be known to the gateway (protocols, names, and addresses are exchanged between systems), and each user must be listed in the central directory unless the autoflow capability overrides this feature. Gateways can also offer an organization the ability to designate which other companies, countries, or even persons, can communicate with their internal gateway users. This is accomplished by means of an access control table listing all the organization's preferences.

Some gateways offer the capability to encrypt messages as they flow out of the gateway to public messaging carriers or across the organization's wide area network. This is a relatively new feature, because in the past e-mail was not used to transmit classified or company-sensitive information.

A trend is emerging toward user encryption keys. The public keys under the Rivest, Shamir, Adleman (RSA) encryption method are stored in a centralized directory such as X.500. Storing this type of information in a directory entry enables each user to encrypt classified and sensitive or proprietary e-mail items reliably and efficiently. Also, the use of the Clipper encryption device, which might become a requirement under federal law, could impact electronic messaging at the system level. If the device is employed, the government will be able to tap into e-mail systems and decrypt messages for law enforcement or national security reasons.

Document Conversion

While many organizations currently use document translators to convert documents, for example, from WordPerfect to Microsoft Word format, a trend is now emerging to offer this capability at the gateway level. This service would perform the operation for all e-mail systems at the gateway and eliminate the costly point-to-point document conversion packages that otherwise must be duplicated throughout an organization.

Access to Utilities

Many gateways offer a centralized approach for access to services such as fax, telex, paper copy delivery, and distributed printing. A gateway provides efficient access to these capabilities, collects statistical information, and simplifies the addressing required to use them. For example, an organization might have hundreds of separate fax machines operating in a free-for-all mode, with no records kept and no centralized listing of fax numbers readily available to all employees. A fax gateway provides this centralization, ensures consistent addressing through a common directory of all fax numbers and their locations, and compiles logs of called numbers, numbers of pages sent, and so on. Users save time by sending faxes directly from their private workstations and receive delivery confirmation, including time of delivery.

PART TWO

Getting Electronic Systems to Work Together

Chapter Five

Connecting to Value Added Networks

Public service providers, more commonly known as *value added networks (VANs)*, operate in the public domain as common carriers of electronic information. Analogous to the telephone network, VANs such as MCI, AT&T, and Sprint, offer a set of services to subscribers for a fee. To become a subscriber, a user establishes an account with a particular VAN and is then able to exchange messages with other subscribers on the same VAN. In this manner, VANs enable different organizations to intercommunicate and share information.

Over the last few years more and more VANs, in addition to offering their intercompany message delivery services, began offering e-mail system interconnection services. In this way, VANs perform the role of e-mail gateways, which are discussed in Chapter 4, "An Introduction to E-mail Gateways." This allows a company to interconnect its various e-mail systems such as cc:Mail, WordPerfect Office, and PROFS to the VAN that then does the system integration by handling all message translations. The company saves on the capital investment necessary to install multiple gateways to interconnect disparate e-mail systems.

All VANs provide a common set of basic e-mail services. Each VAN's particular service offerings are in a constant state of flux. These services and their costs are changing rapidly as providers anticipate current and future markets. This chapter discusses features and functions common to all VANS, including:

- Expected normal functionality
- Types of gateways offered
- Addressing unique to each VAN
- Criteria for selecting a VAN

Overview

Initially there were two types of VANs: those that started by supporting *Electronic Data Interchange (EDI)*, the electronic exchange of business documents between companies, and those that started by supporting e-mail. In the last few years, many of the EDI VANs have added e-mail as one of their products, and e-mail VANs have added EDI to their products. Most VANs understand the synergy of EDI and e-mail in intercompany communications. In keeping with this trend, some current online services have expanded their e-mail services to groups outside of their online service subscriber base. For example, both CompuServe and America Online now offer access to the Internet for the exchange of SMTP e-mail. CompuServe has aggressively marketed their MHS hub concept so that Novell MHS users can attach their networks through the CompuServe MHS hub to communicate with X.400, other MHS sites, the Internet, and other CompuServe users.

Online Services and VANs

Historically, VANs and online services positioned themselves in two very distinct markets. Online services captured mostly small business and individuals, while VANs concentrated on garnering the message traffic of large corporations, institutions, and the federal government. Online services are not regulated common carriers as are VANs. They set their own fee schedules, respond to market demands faster, and are not constrained in their product mix. Companies like CompuServe, America Online, Dow Jones Services, Prodigy, and GEnie are probably the best known and offer a wide variety of services in addition to messaging:

- *Data query*   Access to a large number of commercial databases, containing such information as subscriber addresses, book reviews, new product releases, stock market quotes, stock market buys and sells, and weather reports.

- *Forums*   Electronic discussions, accessible by subscribers, on common issues and ideas. Also known as *bulletin boards,* these forums are often used to answer technical support questions for products such as HP, DEC, Lotus, and Microsoft.

- *Online magazines*   Electronic versions of magazines, such as *Time* and *MacWeek.*

- *Online stores*   Electronic shopping malls for browsing or purchasing various and sundry products and services.

Online services often use their own set of messaging interfaces and protocols. These protocols started out being terminal interfaces and have expanded to support PCs both on and offline. Online services act in part as message stores, or in-boxes, with PCs as the user interface to the messaging and directory systems. Now, they are aggressively implementing interconnections with other networks through X.400 and Internet gateways. These connections often start with basic e-mail and expand to include file transfer access and directory lookups. This is especially the case with Internet connections.

Gateways to VANs

Gateways connecting to service providers are marketed by four sources:

- The service providers themselves
- Major e-mail vendors, which are discussed in Chapter 6, "Description of Private E-mail Systems"
- E-mail integrators
- Small, third-party programming houses

All VANs offer one or more gateways to which e-mail systems can connect. The four most often used are SNADS, Global MHS, SMTP, and X.400, discussed in detail in Chapters 9 through 12. Recently, many VANs, lead by CompuServe, started offering communication hubs for products such as MHS, cc:Mail, and Microsoft Mail, in addition to their standard gateways.

In this configuration, a VAN acts like another gateway connecting different e-mail systems. For example, Microsoft Mail can connect to MCI or PROFS, which can connect to AT&T EasyLink Services. The VAN acts as a message switch to other systems, translating messages, addresses, and attachments as necessary to facilitate intercommunications. It attaches, usually through a dial-up communications facility, and exchanges mail without any additional gateway being necessary.

Additionally, many VANs still offer proprietary gateways to their services, that may or may not offer functionality over and above that offered for X.400 and Internet connections.

X.400 Gateways

Most VANs support the X.400 standard, so purchasing an X.400 gateway gives a company the capability to connect to the greatest possible number of VANs. Prior to 1989, none of the VANs were connected using X.400. This represented a serious interoperability problem. The Aerospace Industries Association enlisted the assistance of seven VANs, requesting them to connect using X.400, so that work on defense contracts could continue when business partners were using different e-mail systems and different VANs.

If a VAN does not currently support X.400, chances are it will in the near future, in order to maintain its competitive edge in this dynamic electronic marketplace.

The X.400 standard and its protocols are discussed in detail in Chapter 12, "X.400 Interpersonal Messaging."

Proprietary Gateways

Several proprietary gateways exist, and the most frequently used are Novell MHS and SNADS. VANs and online services have begun to directly support proprietary message transports such as cc:Mail, Lotus Notes, and Microsoft Mail, without any intermediate gateways being needed. CompuServe has been the most aggressive in this area, offering hubs to Notes, Novell MHS, and cc:Mail. Other VANs, such as MCI and AT&T, are negotiating agreements with vendors such as Lotus and Microsoft to offer these types of direct product connection services.

These hub services are differentiated from normal gateway services in that no additional gateway is required for the LAN. From the LAN e-mail administrator's point of view, it is just as if the connection is to another domain of the same type of e-mail product currently being used.

Internet Gateways

The largest e-mail network in the world is the Internet. It spans the globe, has thousands of networks attached to it, and serves millions of users in over 50 nations. It started in the 1970s as a United States Department of Defense network, called Arpanet, that linked universities doing government research. The Internet currently services commercial, government, and academic interests. Its fastest growing segment is in the commercial area.

Every VAN and online service offers, or will offer in the future, access to the Internet, though they do not always offer Internet connections in the same manner in which they offer X.400 connections. However, that is rapidly changing and, in the near future, a VAN or online service that does not offer access to both will be at a disadvantage in the market.

The Internet and its protocols are discussed in Chapter 11, "SMTP E-mail Services."

Other Services

In addition to their basic messaging services, many VANs offer capabilities similar to those of online services, including the following:

Database queries
E-mail document conversion
Restricted groups
EDI and e-mail
EDI document conversion
Forms and form response systems
Directory lookups
PC and terminal support
Support for mobile users
Direct support for Microsoft Mail Remote and cc:Mail Remote

VANs Offering Both EDI and E-mail Services

An important issue for a company selecting a VAN is whether the VAN supports both EDI and e-mail. As EDI becomes more prevalent across the U.S., e-mail traffic will also expand. In one of its studies, the GartnerGroup found that an EDI transaction, such as a purchase order sent electronically to another company, results in as many as nine supporting e-mail messages.

For example, a purchasing agent from company A sends a purchase order (X12 transaction Number 850) to company B. Company B responds with a purchase order acknowledgment (X12 Number 855) indicating that the order is accepted but item three has been backordered and will be shipped on a specified date.

Normally, this exchange is supported by several phone calls between company A's purchasing agents and company B's sales people to resolve any issues. E-mail is a more effective means for communication between the purchasing agents and the sales force. Increasingly, intercompany e-mail is being driven by this need to support EDI transactions. If the same VAN is used for e-mail and EDI, a company may get larger volume discounts, reduce network management overheads, and present a common functionality set to EDI and e-mail users.

Vendors

Table 5-1 lists the current VAN vendors in North America. They are divided into four categories: e-mail VANs, online services, mobile services, and Internet access services.

Addressing Across the Major Public Systems

Four types of addressing are generally used across VANs: EDI, X.400, Internet, and the VAN's own proprietary type. EDI may use X.400-type addresses, although special address formats are usually the norm. Since the focus of this book is e-mail, the three relevant types are X.400, Internet, and VAN proprietary.

E-mail Interconnect Service Providers	
AT&T EasyLink Services	(800) 242-6005
Advantis	(800) 284-5849
Electronic Mail, Inc.	(818) 403-1999
GE Information Services	(800) 433-3683
MCI	(800) 872-7654
Sprint	(816) 854 2157
Infonet Services Corporation	(310) 335-2600
DA Systems	(408) 559-7434
WilTek, Inc.	(203) 853-7400
Online Service Providers	
CompuServe, Inc.	(614) 457-8600
America Online	(800) 827-6364
Prodigy	(914) 993-8176
GEnie	(800) 638-9636
Mobile Service Providers	
RadioMail	(415) 572-6000
EMBARC	(407) 364-2000
Inmedia Infomatic, Inc.	(514) 397-9747
Ericsson GE Mobile Communications	(201) 265-6600
Internet Service Providers	
Advanced Networks & Services	(313) 663-7610
Performance Systems International	(808) 956-3499
UUNET	(703) 204-8000
SprintLink	(703) 904-2156

Table 5-1. *Value Added Network Vendors*

X.400 Addressing

X.400 is discussed in Chapter 12, "X.400 Interpersonal Messaging." In addition, you can find detailed information on addressing issues between VANS in the Electronic Messaging Association's *X.400 Reference Guide* (Arlington, VA, EMA, 1994). This guide lists all the major service providers and their own rendition of the standard X.400 address. The use of X.400 addressing across systems remains an issue and often requires an understanding of addressing structure and end system requirements. Even subtle variations, such as what is really meant by the "initials" field, can cause addresses to become unresolvable.

Internet Addressing

Internet addressing usually causes few problems between systems and VANs. It is easy to enter into the e-mail system and is widely supported by all VANs and e-mail systems. The Internet address structure is discussed in Chapter 11, "SMTP E-mail Services."

Proprietary Addressing

VANs all have proprietary addressing in place, in addition to their Internet and X.400 addressing schemes. Table 5-2 shows a comparison of different e-mail system addresses. The VAN-specific addressing formats are still in use for two reasons: the historic user knows the formats and how to use them, and the formats support the addressing required to implement the new hub functionality for mail. The hub functionality is what supports the direct connection of cc:Mail, Microsoft Mail, and other proprietary mail systems to the VANs without the need for gateways.

VAN, Protocol, or E-mail Service	Sample Address
X.400	c=us;a=attmail;p=abctool;g=randy;g=smith
MCIMail	754,3456 or BCASEY
CompuServe	76550,1444 or MHS:Julie@ABCTool
America Online	nsmith34@aol.com or NSmith34
Internet	randy_smith@sales.abctool.com
IBM SNADS	SYSNAME (FJSU84VU)
Novell Global MHS	Randy Smith@Sales.ABCTool
cc:Mail	Randy Smith at ABCTool
Microsoft Mail	RandySmith@Finance@ABCTool
ALL-IN-1	Randy Smith AT A1 at ABCTool
HP DeskManager	Randy Smith/Corp/ap
Emc^2/TAO	Randy.Smith

Table 5-2.   *E-mail Address Comparison*

Criteria for Selecting a VAN

The complexities of integrating e-mail systems with a VAN are great. Even though there are solid standards in place for the X.400, Internet, NetWare Global MHS, and SNADS areas, and interoperability test suites are available from the Corporation for Open Systems (COS), detailed interoperability testing to verify the implementation of the e-mail gateways does not always take place.

In addition, there is no standard service level statement for other private e-mail systems that will be attached through the VAN. This means that services such as delivery reports, timely delivery, IPMIdentifier, and so on, may not be supported across all systems and interconnected VANs.

The Electronic Messaging Association (EMA) and the European Electronic Mail Association (EEMA) are working to standardize the service levels of global e-mail. Both have just released a proposed set of service level parameters and services for e-mail systems to follow. The EEMA paper is called, "PRMD Operational Guidelines" (Worcestershire, England, EEMA, 1994), and the EMA paper is called, "Requirements and General Accepted Practices for Operating a Production PRMD" (Arlington, VA, EMA, 1994). These papers address standard operating guidelines for all organizations implementing open e-mail systems to follow in order to provide an acceptable level of service to other organizations with whom they communicate.

VANs are in a highly competitive environment and are continually adding new services to their offerings. Appendix A, "Value Added Network RFP/RFI Technical Skeleton," is a technical outline of a Request for Proposal/Request for Information (RFI/RFP) pertaining to the selection of a VAN. The outline focuses on establishing services and service levels, and is intended for an organization anticipating a sizable amount of e-mail traffic over a period of years that wishes to implement a VAN connection. For such an organization to use the format for a small installation would be overkill. However, the outline identifies general areas useful for anyone choosing a VAN.

Summary

Choosing a VAN and connecting to it requires verifying its services and service levels. VAN vendors are in a competitive market and are always adding functionality and services to their products. The only way to see if those meet the required cost structure and services needed by your company is to do either an RFP or an RFI. The latter is often preferred, since it collects the same amount of technical data as the RFP, while requiring less staffing and time.

Be aware that once you choose a VAN, you may not be able to easily change it later. The VAN address is part of your user's e-mail X.400 address, and to change VANs means all e-mail users registered for VAN use must have their X.400 addresses changed. Those address changes must then be communicated to all your trading partners.

Chapter Six

Description of Private E-mail Systems

This chapter discusses the eleven e-mail systems with the largest number of users today in corporations, institutions, and government entities. The e-mail systems are listed and discussed in alphabetical order by vendor name, not by subscriber base or market share:

- Banyan's BeyondMail
- CE Software's QuickMail
- DaVinci's DaVinci eMail
- Digital Equipment Corporation's ALL-IN-1
- Fischer International's Emc^2/TAO
- Hewlett-Packard's HP Open DeskManager
- IBM's OfficeVision/VM (formerly named PROFS)
- IBM's OfficeVision 400
- Lotus Development Corporation's cc:Mail
- Lotus Development Corporation's Lotus Notes
- Microsoft Corporation's Microsoft Mail

For each e-mail package, the following categories are discussed in detail:

- Product contact information
- Product name and version
- Platforms, operating systems, and network specifications
- Message user agent
- Product directory structure
- Addressing structure
- Message transport system
- Message structure
- Message store
- Gateways supported

By organizing the information in this manner, you can easily compare the different product architectures and the components that facilitate or hinder the interconnection of e-mail systems. These components are message formats, directory formats, directory update methodologies, and message transport mechanisms. In addition, along with Chapters 2, 4, and 7, this chapter forms the foundation for understanding how to create a stable multivendor e-mail environment.

Banyan Systems, Inc. BeyondMail

BeyondMail is a feature-rich, LAN-based e-mail system. Beyond Incorporated was one of the first e-mail vendors to introduce such innovations as rules-based filtering. Rules-based filtering enables a user to more effectively manage routine tasks, including automatic reminders, forwarding of user-specified messages, and integrated forms design and routing capabilities. Figure 6-1 illustrates BeyondMail's architecture, highlighting the computer hardware platforms it supports (DOS, Windows, UNIX) at the top of the drawing, and the transport system it uses at the bottom.

Product Contact Information

Vendor: Banyan Systems, Inc.
Address: 120 Flanders Road
              Westboro, MA 01581
Phone: (508) 898-1000

Product and Version: BeyondMail Release 2.0

BeyondMail for MHS is an easy-to-use, full-featured e-mail system. It comes complete with convenient editing, foldering, and file attachment features. BeyondMail supports scalable address books, message encryption, and automated directory propagation. It can automatically file messages in folders, forward them, or respond to them.

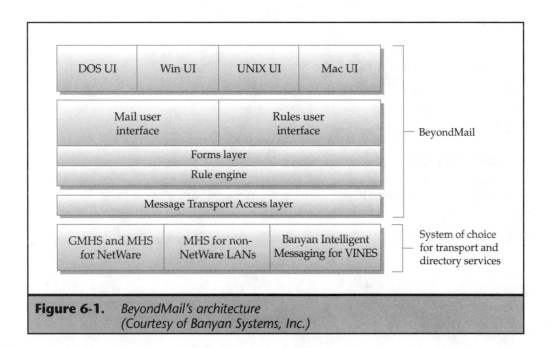

Figure 6-1.   *BeyondMail's architecture*
(Courtesy of Banyan Systems, Inc.)

Additional products include Forms Designer and BeyondRules. The BeyondRules scripting language may be used to develop workflow applications. Some of its important features are:

- Routing slips
- Database access for forms and rules through Q+E drivers
- Watermark Explore edition for image-enabled mail
- NetWare Global MHS support
- Lotus Notes 3.0 Alternative Mailer support
- Support for Outside/In Views for various text and graphics formats
- Conversion of messages to drafts for editing and saving
- Reply to, forward, carbon copies (CC:), and blind carbon copies (BCC)

Editor

BeyondMail supports spell checking, addressing using a search capability, and Rich Text Format (RTF). RTF permits the use of different fonts, font sizes, and color. BeyondMail supports Object Linking and Embedding (OLE), which enables users to insert editable objects created in other Windows applications into the message. For example, users can add spreadsheets, graphics, and other files. BeyondMail also features "drag and drop" capability so you can simply use the mouse to manipulate objects and deposit messages in a folder.

Security

BeyondMail supports these standard security features:

- Original text of forwarded message may not be edited
- Password-protected mailboxes
- Encrypted mailboxes
- Password-protected messages

Outside/In Views

BeyondMail supports the use of Outside/In views, which provides the display and printing of the following types of document text, spreadsheet, database, and graphics formats:

Ami/Ami Pro	Microsoft Word for DOS
ASCII text	Microsoft Word for Windows
.BMP	Paradox
dBASE	.PCX
Harvard Graphics	WordPerfect for DOS
IBM Revisable Form Text	WordPerfect for Windows
Lotus 1-2-3	.ZIP

Intelligent Features

MailMinder is a personal productivity feature that helps you organize mail messages more efficiently. It lets you set up rules that automatically sort all messages into predetermined folders and auto-forward specific types of messages to designated users. The system also includes an "auto-tickler" function to remind you of important upcoming events.

Routing slip functionality lets you route messages to a group of individuals in a predefined, sequential order. You can use this feature to process budgets, expense reimbursement requests, travel requests, and scanned articles and images.

BeyondRules

BeyondRules is a scripting language that can be used to create message-based workflow solutions to help users easily access and share information. BeyondRules can be combined with custom forms to develop applications that automate existing paper-based processes.

BeyondRules supports mail-enabled access to databases through Q+E drivers, which allow access to SQL databases such as dBase, FoxPro, Oracle, Paradox, and Sybase.

Windows Features

BeyondMail supports Windows' Dynamic Data Exchange (DDE) protocol so that users can create and send messages from within other Windows applications. You may attach a file to the message or drag and drop part of the file into the message. BeyondMail supports clients in two environments: DOS and Windows. Also, Novell MHS and Global MHS can supply directory services and mail transport for NetWare; MHS can provide the same for non-NetWare LANs and Banyan Enterprise Network Services (ENS).

OLE support for both client and server lets BeyondMail interact with other Windows applications. The multiple document interface with the savable desktop layout allows you to keep and view more than one window on the screen as well as refer to multiple messages.

Calendaring

BeyondMail's scheduling and calendaring integration capability is provided by CaLANdar, a third-party product by Microsystems Software, Inc.

Platforms, Operating Systems, and Network Specifications

BeyondMail supports all current computer hardware platforms and operating systems, including DOS, Windows, UNIX, and Macintosh. With this level of support, users can exchange messages across multiple platforms, and run workflow applications across them.

The system supports three LAN network options: Banyan Intelligent Messaging for VINES, Global MHS, and MHS for NetWare 4.0. In addition, for non-NetWare LANs, BeyondMail lets users run the e-mail application over any PC-based LAN.

Platforms:	DOS, Windows, UNIX, and Macintosh
Operating systems:	DOS, Windows, UNIX, and Macintosh
Networks:	LAN Manager, NetWare, and VINES
Mail transports:	Banyan Intelligent Messaging, Global MHS, and MHS

Message User Agent

BeyondMail's user interface is graphical and icon-oriented. It employs pull-down menus, lists, and dialog boxes to facilitate access to the product's various features. Figure 6-2 illustrates the graphical user interface of BeyondMail's Windows application. As shown, it includes the standard toolbar for submenu lists and icons for accessing the in-box, printing, attaching files, sending messages, spell checking, and changing fonts, among others.

Product Directory Structure

BeyondMail's directory system is based on Novell's MHS. The product may also use Banyon StreetTalk as the directory component. Novell MHS is discussed in Chapter 10, "Novell NetWare Global Message Handling Service." Banyan's StreetTalk directory is discussed here.

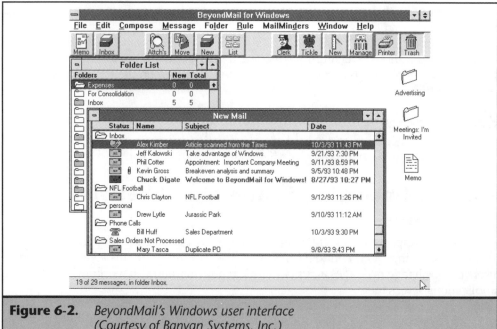

Figure 6-2.   *BeyondMail's Windows user interface
(Courtesy of Banyan Systems, Inc.)*

The Directory Record

Banyan's StreetTalk directory implements the object/attribute structure. Each object has a three-part name and associated attributes. The name is similar to the X.500 distinguished name discussed in Chapter 8, "X.500 Directory Systems." The three-part name uses the format:

Item_Name@Group_Name@Org_Name

Physical address locations, such as the X.500 directory system agent discussed, are described by the pair:

Group_Name@Org_Name

Each of the pairs describes one of the distributed database engines that compose the StreetTalk directory. The StreetTalk directory includes much more than just e-mail addresses; it also stores all the parameter, routing, and user demographic data used by Banyan VINES or other enterprise network services.

Each object may have an almost unlimited number of attributes, such as the user name, nickname, mail service, server, location, postal address, organizational role, phone number, e-mail address, and a picture of the employee.

These attributes do not exist until they are defined by the administrator and therefore do not take up space until added.

Mechanism of Directory Update

BeyondMail employs the directory synchronization services of either Global MHS or StreetTalk. Global MHS is discussed in Chapter 10, "Novell NetWare Global Message Handling Service."

Banyan's StreetTalk service is composed of two closely linked services: StreetTalk and StreetTalk directory assistance. These two components work together to give the user a full-function set of directory services.

StreetTalk is a fully distributed, partially replicated, network directory. StreetTalk uses the replicated information to route directory queries and to maintain a single view of all groups and organizations on a network. StreetTalk replicates changed information using a variety of algorithms. These replication algorithms ensure 99.99% accuracy at any point in time, while increasing network traffic by less than 2%.

StreetTalk Directory Assistance is a partially distributed, fully replicated directory that uses a multikey ISAM database to integrate StreetTalk and non-StreetTalk objects for user queries. This client-server database is synchronized periodically as specified by the administrator, because StreetTalk objects change within two network hops, and automatically every night.

Addressing Structure

BeyondMail uses the Novell or the Banyan addressing schemes, depending on which transport (Novell or Banyan) supports the application. Novell addressing is described in more detail in Chapter 10, "Novell NetWare Global Message Handling Service."

Banyan's StreetTalk names are fully qualified; you must specify only as much of the name as necessary to uniquely identify the recipient. For example, you may address Steve, who works in the same department, simply by using the address:

Steve

Address Steve in Manufacturing by using:

Steve@Manufacturing

and Steve in Sales in Canada with:

Steve@Sales@Canada

Banyan StreetTalk supports several different types of addressing formats:

■ StreetTalk names use the form:

John@Sales@ABCTools

■ List names, which are distribution lists, use the form:

SalesPersonnel@Sales@ABCTools

■ Nicknames, which are short form user names, locally maintained by one user

■ Patterns use the form:

*@Sales@ABCTools

The pattern in this example addresses the message to all users in Sales.

■ Remote electronic mail addresses (REMAs) use of the form:

Gateway_name@Sales@ABCTools [*name_on_other_side_of_gateway*]

Message Transport System

BeyondMail uses one of two transports: Novell MHS or Banyan Intelligent Messaging. MHS is discussed in detail in Chapter 10, "Novell NetWare Global Message Handling Service." Banyan Intelligent Messaging is discussed here.

Banyan Intelligent Messaging has message transport agent (MTA) software that runs on servers. The MTA moves mail among local and remote mailboxes, and gateways. The MTA transfers messages and usually places them in first-in-first-out (FIFO) queues. Several queues exist: Accept, Send, Resolve0, Resolve1, Resolve2, Transfer0, Transfer1, and Transfer2.

When a user sends a message, it is first placed in the Send queue and then moved to the Resolve0 queue. Messages delivered to a server from another network address, on the other hand, are placed in the Accept queue and later moved to Resolve0 queue.

The Resolve0, Resolve1, and Resolve2 queues are used to resolve, or find, the StreetTalk names in the addresses. The MTA spends a limited amount of time trying to resolve the address of each message in the queue. If the message's addresses are completely resolved by the MTA, the message is moved to the corresponding Transfer0 queue. Similarly, Resolve0, Resolve1 and Resolve2 messages are moved to the Transfer0 message queue after their addresses have been completely resolved.

After the allocated time for an individual message expires, the MTA moves on to the next one in the queue. Low and normal priority messages, whose addresses are not resolved on the MTA's first pass through Resolve0, are moved to Resolve1. High priority messages remain in Resolve0. Messages in Resolve1, which have had their addresses resolved after two tries, are moved to Resolve2.

Messages with resolved addresses are stored in Transfer0 queue. The messages are evaluated based on the network cost to transfer the messages. Low cost messages are moved to Transfer1; high cost messages are moved to Transfer2. The cost tables are maintained by the administrators. This multithreaded transport system is shown in Figure 6-3.

If the message is destined for a user on the same server, the MTA delivers the message to the user's in-box. If the user is on a different server, the message is delivered to the Accept queue on the destination server.

The MTA software automatically compresses and decompresses messages that are 5,000 bytes or larger, including attachments. Also, the MTA uses checkpoints to transfer messages over links. If a message is only partially transferred, just the part from the last checkpoint must be re-sent—not the whole message. This is a major benefit for networks that are unreliable by nature.

Message Structure

The BeyondMail message structure can be either Novell SMF or Banyan Intelligent Messaging. Novell SMF-70 and SMF-71 structures are discussed in detail in Chapter 10, "Novell NetWare Global Message Handling Service." Banyan Intelligent Messaging is discussed here.

The Enterprise Messaging Services (EMS) Intelligent Messaging release 3's message structure is defined by a set of Banyan Application Program Interfaces (APIs). These new APIs and fields are backward-compatible with VINES 5.50 and EMS 1.10. The fields that make up the EMS envelope are a combination of the MAPI and X.400 1988 data structures.

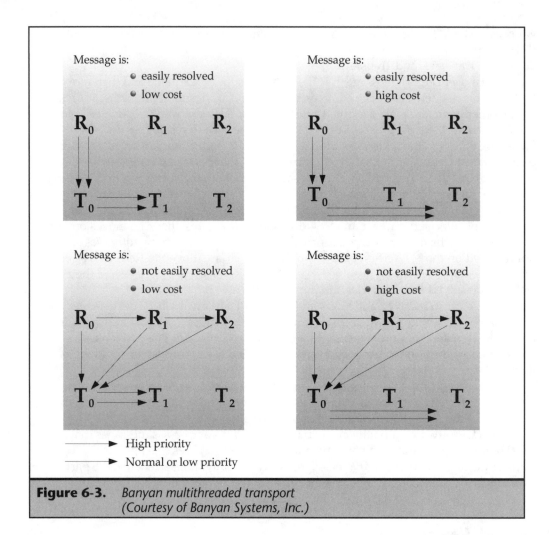

Figure 6-3. *Banyan multithreaded transport
(Courtesy of Banyan Systems, Inc.)*

Envelope Fields

EMS uses a set of Banyan messaging APIs to describe the message. It contains a series of predefined envelope and body structure information fields, as listed here. Other fields may be defined by application developers to provide additional functionality.

Field Name	Field Contents	Field Type
VNS_KEYWORD	A user defined keyword with a maximum size of 31; a message may have up to 10 keywords	Character string

Field Name	Field Contents	Field Type
VNS_SMALL_ICON	A 16 x 16 x 16 Windows icon. The mail service does not use this field, but the client application may	Bit string
VNS_ICON	A 32 x 32 x 16 Windows icon. The mail service does not use this field, but the client application may	Bit string
VNS_CREATION_DATE	Creation date of the message	
VNS_CREATE_GMT	gmt_seconds	
VNS_SEND_DATE	Send date of the message	
VNS_SEND_GMT	gmt_seconds	
VNS_EXPIRE_DATE	Expiration date of the message	
VNS_DEFER_DATE	Defer date of the message (the service waits until this date to deliver the message)	
VNS_READ_DATE	Date the message is initially read (the service sets this field when the status changes from VNS_UNREAD to VNS_READ)	
VNS_RESPOND_BY	A requested respond by date (the service does not enforce this field)	
VNS_PRIORITY	Delivery priority of the message (values are VNS_HIGH, VNS_LOW, and VNS_NORMAL)	Character
VNS_IMPORTANCE	User's assessment of the importance of the message (values are VNS_URGENT, VNS_IMPORTANT, and VNS_ROUTINE)	Character
VNS_SENSITIVITY	Sensitivity of the message (values are VNS_ROUTINE, VNS_PRIVATE, VNS_PERSONAL, and VNS_CONFIDENTIAL)	Character
VNS_CERTIFIED	Indication of whether the sender wants notification when the recipient accesses the message (default is VNS_NO)	Character
VNS_DELIVERY_RECEIPT	Indication of whether the sender wants notification of successful delivery of the message (default is VNS_NO)	Character
VNS_NON_DELIVERY _CONTENTS	Indication of whether the sender wants the original message included with any nondelivery receipt (default is VNS_YES)	Character
VNS_AUTO_FORWARD	This field is set if the message should automatically be forwarded; for example, if the user is on vacation. This field is not currently supported	Boolean
VNS_NESTED_TYPE	Identification of nested message as VNS_FORWARD_MSG or VNS_REPLY_MSG	Character
VNS_NESTED_DEPTH	Current nesting depth of the message	Word
VNS_TRACE_INFO	Trace information set by mail services (this field is read-only)	Character
VNS_STATUS	User access status of the message (values are VNS_READ, VNS_UNREAD, VNS_SENT, and VNS_UNSENT)	Character

Field Name	Field Contents	Field Type
VNS_INCOMPLETE	This field is set if the message contains incomplete data, usually because information was lost during conversion (default value is VNS_FALSE)	Boolean
VNS_MESSAGE_TYPE	Message type (VNS_VINES_MAIL) should be used by vendors preparing a standard interpersonal mail message; VNS_STATUS_MSG and VNS_UNDELIV_MSG are reserved for messages prepared by the VINES Mail service	Character
VNS_CONTENT_TYPE	Content type of the MAINBODY part of the message (values are FREETXT1 and FREETXT2)	Character
VNS_APP_TYPE	Creator of the message if VNS_MESSAGE_TYPE is set to VNS_APP_SPECIFIC (should correspond to a specific mail client vendor)	

The following message creator types are predefined; other types will be defined as indicated by developers.

VNS_BEYOND
VNS_CMS
VNS_LANSHARK
VNS_LOTUS
VNS_MACSOFT
VNS_MICROSOFT

VNS_NETPRO
VNS_REACH
VNS_SOFTSWITCH
VNS_STARNINE
VNS_ZOOMIT

■ VNS_MESSAGE_ID   This field contains a unique message ID provided by the service. The field type is character.

■ VNS_CONVERSATION_ID   This field contains the VNS_MESSAGE_ID of the original message that initiated the reply and forwarded the message. The field type is a bit string.

■ VNS_ENVELOPE   This field contains the pointer to the complete, older-type envelope. The field type is value.

■ VNS_BODYPART_INFO   This field value contains a pointer to the construct body part. The field type is value.

Body Structure Fields

The body of a typical message—the written text—is structured as a series of fields. A message containing text and one attachment, such as a graphics file, would have two sets of fields, one for each body part.

■ *Present*   This field indicates that the rest of this structure has meaningful values.

■ *Bodypartid*   This field indicates the unique body part identifier.

- *contentType*   This field has two predefined values: VNS_VINESMAIL and VNS_VINES_MACMAIL.

- *contentSubType*   This field contains values that further qualify those in contentType:

ContentType	Values
VNS_VINES_MACMAIL	VNS_MACFILE and VNS_MACDATA
VNS_VINESMAIL	VNS_NODATA, VNS_TEXT, VNS_UNWRAPPED_TEXT, VNS_BINARY, VNS_APPLESINGLE, VNS_MS_RTF, VNS_MAC_STYLED_TEXT, VNS_PCX_FAX, VNS_G3_FAX, VNS_G4_FAX, VNS_DIB, VNS_WINMETA, VNS_OS2META, VNS_TELETEX, VNS_VIDEOTEX, VNS_MAC_PICT, VNS_QUICKTIME, VNS_MAC_FAX, and VNS_MS_OLE

- *label*   This field is a text label for the body part.

- *contentAttributes*   This field contains additional attribute specifications for the message content.

Message Store

BeyondMail uses either the Novell or the Banyan Intelligent Messaging post offices. Messages may be stored locally or on the server. All products may share a common message store on the server.

Gateways Supported

BeyondMail supports a variety of e-mail gateways, depending on whether the message transport is VINES or MHS. It maintains gateways to most of the popular e-mail systems in use today, including both mainframe- and LAN-based systems. Banyan Intelligent Messaging transport supports these gateways:

Banyan Systems' SMTP gateway
BISCOM's Fax gateway
Computer Mail Services' cc:Mail gateway
Computer Mail Services' MCI Mail gateway
Computer Mail Services' MHS gateway
Computer Mail Services' Notes gateway
Computer Mail Services' V-Bridge/MHS gateway
ILAN's ALL-IN-1 gateway
Incognito Software's Intelligent SMTP gateway
MacSoft's cc:Mail gateway

MacSoft's Wang Office gateway
Nettech Systems' CA-eMail+ gateway
Soft*Switch's SNADS gateway
StarNine Technologies' Mail*Link/VINES to CE Software
StarNine Technologies' QuickMail Mac gateway
ZOOMIT's cc:Mail gateway
ZOOMIT's X.400 gateway

If the MHS transport option is selected, BeyondMail supports a similar set of e-mail system gateways for LAN-based systems:

Computer Mail Services' V-Bridge/MHS gateway for VINES
Data Access' MHS gateway for DATA Access OfficeWorks
Enable Software's Higgins to MHS gateway for Higgins
Lotus' cc:Mail gateway MHSlink for cc:Mail
Lotus' MHS gateway for Lotus Notes
Microsoft's Microsoft Mail gateway to MHS for Microsoft Mail for PCs
StarNine Technologies' Mail*Link MHS for CE Software QuickMail
StarNine Technologies' Mail*Link MHS gateway for Microsoft Mail for Macintosh
StarNine Technologies' Mail*Link MHS for Sitka Inbox
Transend Corporation's MHS gateway for Transend CompletE-Mail
WordPerfect's WP Connections MHS gateway for WordPerfect Office

CE Software QuickMail

CE Software is one of the leading developers of LAN-based groupware applications. With over 2.8 million users, CE Software's QuickMail provides messaging, calendaring, and scheduling applications for DOS, Windows, and Macintosh platforms. In the future, the product will support X.400 through the Common Messaging Calls (CMC) standard, proposed by the X.400 API Association. CE Software also recently released a new product called Mail Rules, a personal digital assistant (PDA) e-mail application for Apple's Newton.

Product Contact Information

Vendor: CE Software, Inc.
Address: 1801 Industrial Circle
P.O. Box 65580
West Des Moines, IA 50265
Phone: (515) 224-1995
Fax: (515) 224-4534

Product and Version: QuickMail V2.6 for Macintosh, DOS, and Windows

QuickMail is an easy-to-use e-mail package for PC-compatibles connected to Apple Filing Protocol (AFP)-compatible servers and Macintoshes running over AppleTalk-compatible networks. QuickMail implements the same user interface on DOS, Windows, and Macintosh clients. QuickMail may be launched from within other applications using a hot key or function as a desk accessory on the Macintosh. DOS, Windows, and Macintosh users are presented with a consistent interface and can seamlessly use QuickMail to intercommunicate without gateways.

QuickMail MHS lets Windows and Macintosh users share messages directly with other MHS users on the same Novell server. With QuickMail, you can:

- Attach over 60 documents to a single message in MHS
- Attach 16 documents to a single message for non-MHS transports
- Access gateways to various e-mail services
- Conduct real-time conferencing between clients
- Address messages to individuals or distribution lists
- View, edit, and print documents in their native application
- Unsend a message and retrieve it before it is read
- Track whether messages have been received, read, filed, printed, or deleted
- Record audio notes in messages
- Create custom address books
- Access bulletin boards

QM Forms

QM Forms software lets any user design an unlimited number of custom message forms, such as purchase orders, mileage reimbursement forms, and expense reports.

QM Remote

QM Remote provides remote access for PowerBook users, with the same interface and functionality as for LAN clients. With QM Remote, you can access the same mailbox whether you are in the office or on the road.

Scheduling

QuickMail supports Powercore's cross-platform Network Scheduler 3 for calendaring and scheduling. Network Scheduler 3 is the only group scheduling application that handles multiple mail transports, including cc:Mail, MHS, Microsoft Mail, and QuickMail.

QuickMail AOCE for the Macintosh

QuickMail for Apple's Open Collaborative Environment on PowerTalk networks supports:

- QuickMail access directly from the finder
- AOCE mail-enabled system extensions
- Access to other AOCE services to link to non-Macintosh networks
- QM Forms to automate the workflow process

Platforms, Operating Systems, and Network Specifications

Platforms:	DOS, Windows, and Macintosh
Operating systems:	DOS, Windows, and Macintosh
Networks:	3COM, AppleTalk, Banyan, DEC Pathworks, LAN Manager, LANtastic, and Novell
Mail transports:	AOCE, Global MHS, MHS, Novell MHS 1.5, and proprietary

Message User Agent

The primary advantage of QuickMail's user interface, as shown in Figure 6-4, is that it is common across all platforms: DOS, Windows, and Macintosh. Organizations with a large number of users on different platforms can minimize their training costs because of this consistent interface.

The user interface is graphical and icon-oriented, a feature shared with most LAN-based e-mail systems. Users employ menus, dialog boxes, and "point and click" capability to navigate through the system.

Figure 6-4 depicts a view of the user's in-box with the different types of messages: received, normal, status, receipt, and so on. The icons at the top of the screen perform such functions as sending and printing messages.

Product Directory Structure

The directory of the QuickMail product is composed of two parts: the local address list and the name server. The name server, which runs on a Macintosh, holds the names of the e-mail users who are on other mail centers and/or use different e-mail products. The name server queries its set of mail centers or gateways at an administrator-defined interval and retrieves information about deletions, additions, and changes. Several name servers might serve the same group of mail centers to enhance user address lookup. QuickMail employs the name servers for remote addresses while the gateways use them to reconcile inbound message addresses.

The format of a name server address record is

Last Name	Address #1 (Street address)
First Name	Address #2 (Street address)
QuickMail Address	City
QuickMail Mail Center	State
QuickMail Network Zone	Zip Code
(AppleTalk subnetwork designator)	Country
In/Out Workgroup	Phone #1
In/Out Network Zone	Phone #2
Department (User's department)	

Addressing Structure

The addressing structure is composed of two levels: the user name and the mail center. Addresses of DOS QuickMail users have a maximum eight-character user name followed by a maximum eight-character mail center name. In this case, users' names also act as their mail folder names in the QuickMail file system. For example, a QuickMail DOS user address might be:

RSmith@MailCen

Figure 6-4.   *QuickMail user interface*
(Courtesy of CE Software, Inc.)

Macintosh QuickMail users can have user and mail center names of up to 32 characters. For example, the same user's address might be:

Rich Smith@MailCenter1

The inter-QuickMail gateways (called bridges by QuickMail) translate the DOS and Macintosh addresses. The e-mail users on other mail systems might have an address such as:

Rich Smith@MailGateway_Name

Gateways map the QuickMail addresses to the foreign address strings so the user doesn't have to know other address formats or technologies.

Message Transport System

QuickMail employs a straightforward message transport system, with the flexibility to accommodate the routing of messages within the same server, between different servers, and between different mail systems.

Volume Directory Structure

The server can either be a PC or a Macintosh. Each MailCenter on the server is a main level directory under the root on the PC, or a major folder on the Macintosh platform. User accounts are defined by subdirectories in the DOS environment and subfolders on the Macintosh. When a user sends the message, a pointer is placed in the receiver's directory or folder. Also, the mail log of the sender, which records message status, is updated and an entry is inserted in the destination user's mailbox (in-box).

Message transport in the QuickMail environment is accomplished in four different ways, depending on the destination of the message. The components are slightly different on the DOS and Macintosh servers. However, the general logic is the same and understanding one enables the user to understand both. The following discussion explores three cases on the Macintosh platform:

- The users are on the same server
- The users are on different servers
- The users are using different mail system products

Users on the Same Server   The user creates and sends a message using the client software. The message is given to the QM server software, which verifies its format and addresses. The software then places the message in the sender's user folder, which is a subfolder of the user's MailCenter folder on the server, as shown in Figure 6-5. QM puts a small file pointing to the message in the recipient's user account and

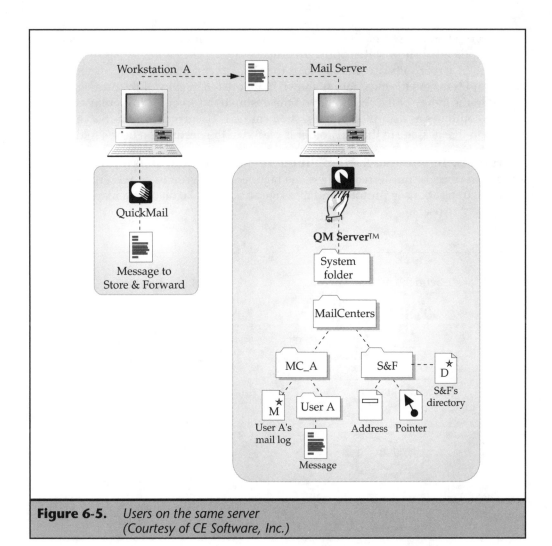

Figure 6-5.   *Users on the same server*
(Courtesy of CE Software, Inc.)

creates an entry in the sender's mail log file. This file is used to track the message status, such as read, delete, unread, and so on. The QM server software then creates an entry in the recipient's MailBox file. (The MailBox file contains both unread and read messages, which are still unfiled.) When the recipient sees the message in his or her in-box (mailbox file) and opens it, the original message appears on the user's screen.

Users on Different Servers   Sending mail to a user on another server involves a store-and-forward process. When the client creates and sends the message, the file that points back to the message is placed in the S&F (store-and-forward) folder with a copy of the address list; the pointer can be placed in the destination user's folder because it does not exist on this server. The Mailbox file in the S&F folder and the sender's log file are updated to show status of the message.

The QM server software then contacts the QM server software on the destination server and transfers the message, as shown in Figure 6-6. A copy of the message is placed in the in-box of the S&F folder on the destination server. At this point, the process continues as described in the section entitled "Users on the Same Server" and a pointer to the message in the S&F folder is placed in the destination user's folder.

Users on Different Mail System Products   Mail between users on different mail system products is transferred in the same manner as between two QM servers. In addition, translations in protocol and message structure take place in a manner that is transparent to users.

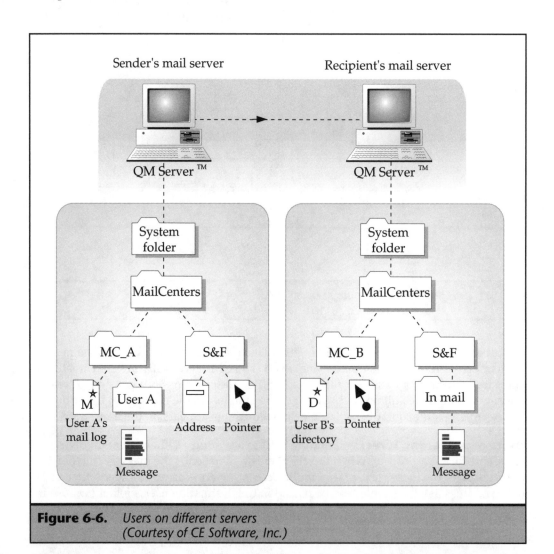

Figure 6-6.   *Users on different servers*
          (Courtesy of CE Software, Inc.)

Message Structure

The QuickMail message is composed of a sequence of fields. Each field has two components: field data and field length. These fields are always located in the same position in the message data; for example, the Subject always follows Return Receipt Request, and precedes Priority. Because a field's position and length are known, it is possible to correctly interpret the information. The message is defined as a structured form; it is not a freeform message as in SMTP. The field encoding is between the freeform, simple SMTP-type message structures, and the complex Abstract Syntax Notation (ASN.1) encoded X.400 messages.

The QuickMail Message comprises seven main sections: Heading, Form Items, Clipboard Attachments, Message Body, Recipient Information, Sender Information, and the Enclosures section.

Heading Section

Heading fields are detailed here:

- Return Receipt Request
- Subject (27 characters in length)
- Priority (one of five levels)
- Send Date
- Been Read Flag

Form Items Section

The Form Items section can hold a maximum of 100 items of the following types:

- Graphics objects, such as rectangles and lines
- Lines
- Check boxes
- Static Text Descriptors
- Edit Text Field Descriptors
- Bitmaps

Clipboard Attachments Section

- Freeform data attached by user

Message Body Section

- Message bodies (a maximum of 32 text bodies of up to 28 kilobytes each)

Recipient Information Section

- Lists of all TO recipients of the message (up to 250)
- Lists of all carbon copy (CC) recipients of the message (up to 250)

Blind carbon copies (BCC) are not part of the message structure; they are handled in a special manner and are not part of this section.

Sender Information Section

- All senders of the message and their addresses

Enclosures Section

- Audio attachment (only one)
- All enclosures (up to 16)

Message Store

The DOS and Macintosh versions both have the normal file cabinet functionality.

Gateways Supported

There are over 100 gateways and bridges that offer support for QuickMail. A few of the most well known include:

- PowerTalk/QM gateway from StarNine
- QM-Link gateway for AppleLink mail systems
- QM-Script gateway for exchanging messages with text-based e-mail services such as GEnie, CompuServe, and Internet

DaVinci eMail

DaVinci, currently integrated with Novell's MHS mail transport and the NetWare operating environment, was designed in an open manner to allow the product to easily support X.400 and other messaging protocols.

DaVinci is committed to conforming to the application programming interface standards of VIM, MAPI, and OCE, as they emerge and become viable in the marketplace.

Product Contact Information

Vendor:	DaVinci Systems Corporation
Address:	4200 Six Forks Road
	Raleigh, NC 27609
Phone:	(919) 881-4320
Fax:	(919) 787-3550

Product and Version: DaVinci eMail V2.5

DaVinci eMail is the leading NetWare MHS-based e-mail application in the world. It is optimized for NetWare but will work on any file-sharing network. DaVinci eMail has over 1.7 million users worldwide across several platforms: DOS, Windows, and Macintosh. It supports networks ranging from those with only a few users to those with several thousand. DaVinci seamlessly connects to NetWare's security and directory services. DaVinci eMail provides a complete menu-driven, graphical user interface (GUI) for easy administration and use. Adding, deleting, and moving users is as easy as dragging and dropping the user's names. You can point and click to perform custom mailbox purging from one screen to save time and disk space.

DaVinci eMail provides native support for Novell's NetWare MHS and Global MHS transport as well as the inherent messaging API, Novell's Standard Messaging Format (SMF). For this reason, no gateways are required to communicate with other electric mail products based on Novell's MHS standard.

Using Novell's Global MHS and the long and short addressing methods, users can communicate with other mail systems through numerous Novell and third-party gateways. These gateways support communications with cc:Mail, CompuServe, IBM OV/VM (PROFS), MCI Mail, Microsoft Mail, and UNIX mail, among others.

Languages

DaVinci eMail supports several languages: American English, British English, French, Hebrew, German, Italian, Kanji, Portuguese, Spanish, and Swedish.

Features

DaVinci eMail supports conversation threading, which allows a user to chronologically view the messages related to a conversation, regardless of where they are filed. It ships with multiple-document viewers that let users view attachments without exiting the e-mail application. It provides shared mailboxes, which can be accessed by multiple users, and a multiuser bulletin board system.

DaVinci eMail Features   The product offers a full-function message editing facility, including search/replace, cut/paste/copy, context-sensitive help, spell checking, and saving of drafts. In addition, DaVinci offers these features:

- File attachments of any type and number
- File attachment views integrated for Ami Pro, Excel, graphics and fax, for both the DOS and Windows versions, Lotus 1-2-3, Microsoft Word, Paradox, and WordPerfect
- Sophisticated message management, such as multilevel folders; drag and drop filing; message-thread searching; searches by date, subject, from, to, and other attributes; message sorting by date, from, to, subject, and priority
- Bulletin boards supported by a menu-driven interface

- Addressing features supported include Novell full names, both public and private address books, mailing lists (distribution lists), 31-character names, customized address templates for gateway addressing, and Global MHS short and long names

- Security ties into NetWare security, including password-protected log in, password-protected individual messages, and encrypted messages

- Over 25 predefined electronic forms such as purchase orders, time sheets, phone messages, and sales orders

- Other Windows features, including voice/sound attachments, toolbar, message browser, Dynamic Data Exchange (DDE), and TrueType fonts

- Pull-down menus with optional mouse support and high-resolution mode in DOS, plus the ability to launch DOS programs

- Administrative features including GUI, synchronization with NetWare Bindery for directory services, automated maintenance on NetWare, self-optimizing databases, optional local message storage, and native NetWare support

Other Related DaVinci Products

DaVinci offers several other products that add functionality to the e-mail application:

- DaVinci CaLANdar v2.5 for DOS and Windows works for both LANs and WANs (wide area networks). CaLANdar supports scheduling of meetings, resources, and people, and recognizes the differences of each. It supports tasks and To Do lists. User-defined security rights allow you to specify who can see your calendar and the level of information may be viewed.

- The DaVinci Assistant implements auto-forwarding, auto-response, and a tickler alarm system. Messages may be automatically routed to a specified user and/or answered with predefined text. Messages may also be posted with delayed delivery for reminders.

- DaVinci eMail Remote offers remote access to PCs not attached to the LAN. Remote uses the same user interface as the LAN-attached eMail product and connects via modem with Personal MHS from Novell.

- DaVinci Wireless eMail based on RAM Mobile Data network, Business Network Ltd.'s everywhere, and Novell MHS mail transport uses the same eMail v2.5 software as LAN-attached devices.

- The Coordinator version II is a workflow, task management system, based on MHS. The Coordinator supports workgroup management by coordinating the actions and commitments of users working together on projects. The product features group scheduling, time management, and automatic notification of pending commitments and deadlines. The Coordinator implements seven message types that provide the basis of information tracking and sorting:

> *Note* Informal freeform text
> *Inform* Convey information
> *Question* Ask for information
> *Offer* Propose action
> *Request* Ask for someone to act
> *Promise* Agree to an action
> *What if* Speculate on an action

Using these seven message types, the conversation management ability, and the group scheduling feature, users can organize and track all communications, commitments, and tasks associated with a group project.

Platforms, Operating Systems, and Network Specifications

Platforms: IBM PC and compatibles, Macintosh Plus or higher

Operating systems: DOS V 3.0 or higher, Windows 3.0 or higher, OS/2 for Windows, and Macintosh System 6.05 or higher

Networks: NetWare 2.*x*, 3.*x*, and 4.*x*, as well as other file-sharing networks, including Banyan, IBM LAN Server, LAN Manager, LANtastic, and NetWare Lite

Mail transports: NetWare Personal MHS v.1.1 or higher, Novell Global MHS or MHS v.1.1 or higher, and PC-LAN MHS v.1.1 or higher

Message User Agent

DaVinci eMail supports three message user agents: DOS, Windows, and Macintosh. The Windows version, shown in Figure 6-7, presents the user with an intuitive, easy-to-navigate graphical interface for the send a message capability.

Product Directory Structure

DaVinci supports both public, systemwide address books and private, user address books. The public address book is maintained by the system administrator. It synchronizes with NetWare Directory Bindery by automatically importing changes, additions, or deletions to the network system. The individual user may access the public address book. The user may also create and customize private address books by entering a nickname, comment, title, and company for each local, private address entry.

Addressing Structure

DaVinci supports both the short addressing schemes used by Novell MHS and the long addressing methods used by Novell's Global MHS (more than eight characters). Global MHS allows users with both long and short names. Long names help differentiate users in larger installations; short names are maintained to ensure

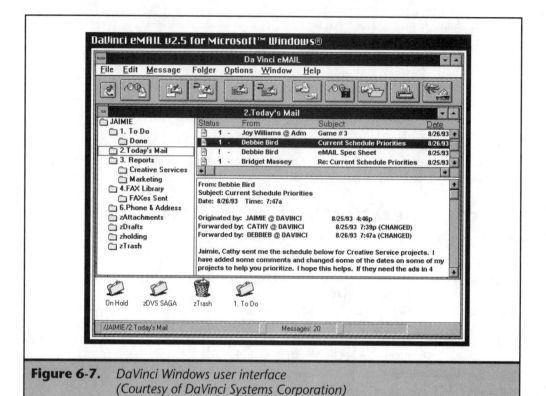

Figure 6-7. *DaVinci Windows user interface (Courtesy of DaVinci Systems Corporation)*

compatibility with non-Global MHS installations. This feature allows DaVinci users to send mail and attachments to users on other front-end e-mail systems and transports.

Message Transport System

DaVinci products are based on the Novell MHS or Global MHS e-mail and directory products.

Message Structure

DaVinci products run on the Novell MHS and Global MHS transport. They also integrate with the NetWare directory structure.

Message Store

The DaVinci eMail product is based on a client-server model. Messages may be stored either on the server or on local disks. The product may use any one of several file-sharing network systems.

Gateways Supported

DaVinci eMail can use many of the MHS gateways available from third-party vendors to access other e-mail systems or network services such as faxing, forms, and paging. The available gateways include, but are not limited to:

AT&T EasyLink	MCI Mail
CompuServe	PROFS
DEC MAILbus/VAX	SNADS
HP DeskManager and OpenMail	Sprint Mail
Internet	UNIX

Digital Equipment Corporation's All-IN-1

Historically, ALL-IN-1 has been DEC's offering in this arena. ALL-IN-1 runs on DEC VAX/OpenVMS machines and uses proprietary message formats, transport protocols, and directory structures. The proprietary structure closely mimics the X.400 and X.500 standards. ALL-IN-1 has an installed base of millions and is the primary focus of this discussion.

In addition to the historic ALL-IN-1, DEC offers full-function X.400 1988, UNIX SMTP, X.500 1988 (with proprietary 1993-like extensions), and X.500 Directory Synchronizer. DEC also provides MAILbus Postmaster for LAN/WAN and MAILbus 400 Message Store for complete integration of user agents such as cc:Mail and Microsoft Mail into the DEC messaging architecture. In sum, DEC is a market leader in the implementation of full-functioned, standards-based electronic messaging systems.

As depicted in Figure 6-8, DEC offers an e-mail backbone called MAILbus that provides connectivity from its ALL-IN-1 product to other popular mainframe and LAN-based e-mail systems. MAILbus is continually updated to provide interoperability with emerging e-mail systems and international standards.

Product Contact Information

Vendor:	Digital Equipment Corporation
Address:	110 Spit Brook Road
	Nashua, NH 03062
Phone:	(800) DIGITAL

Product and Version: ALL-IN-1

ALL-IN-1 Integrated Office System server and ALL-IN-1 Core Services for OpenVMS VAX Version 3.0 are the key Digital offerings.

ALL-IN-1 Integrated Office System (IOS) server for OpenVMS is a customizable, menu-oriented software product that provides generic office applications, a facility for integrating other business-oriented applications, and ALL-IN-1 IOS base services to

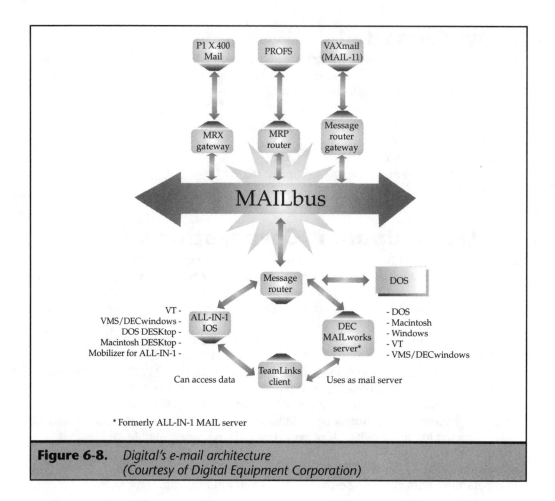

Figure 6-8. *Digital's e-mail architecture
(Courtesy of Digital Equipment Corporation)*

TeamLinks clients. TeamLinks supports clients such as Microsoft Windows and Apple Macintosh desktops. These clients use the basic ALL-IN-1 e-mail, directory, and foldering (message store) facilities.

ALL-IN-1 Core Services for OpenVMS VAX is the full ALL-IN-1 Integrated Office System server product minus the license to use the WPS-PLUS word processing editor and the Time Management (TM) subsystem.

ALL-IN-1 offers facilities for mixing groups of Virtual Terminals (VTs), Windows, and Macintosh (TeamLinks) users to access applications in a consistent manner. User documents are stored in the file cabinet, from which they are accessed by the office applications, such as ALL-IN-1 and TeamLinks.

TeamLinks is a messaging and file-sharing client-server software product that executes on several LAN platforms. TeamLinks makes it possible for Windows and Macintosh clients to participate in the ALL-IN-1 messaging and file cabinet system.

The ALL-IN-1 product set consists of:

ALL-IN-1 DESKtop Server for OpenVMS VAX
ALL-IN-1 Group Conferencing
ALL-IN-1 Office Applications
ALL-IN-1 TeamLinks Connection V2.0
CDA Document Converter Library
Message Router OpenVMS gateway
VAX Forms Management Services (FMS) and FMS Forms Language Translator

File Cabinet/File Cabinet Management

The ALL-IN-1 file cabinet is similar in function to a standard office filing cabinet in that it has drawers of folders that contain documents. Folders are stored in alphabetical order and the documents within them appear in order by date and time of creation. The most recent document is listed first (except in the INBOX, where it appears last).

All previously created folders are stored in the MAIN drawer, along with the messaging folders: INBOX, OUTBOX, CREATED, READ. Users can create other drawers, which can be shared or private.

The file cabinet provides access to documents through a common mechanism, regardless of their format. Support is provided for ASCII, DX, and WPS-PLUS formats.

Shared Filing

Users may create shared drawers to allow others to read and edit shared documents. The drawer owner may specify which users can access it and the operations they can perform. Certain actions are prohibited while a user is modifying a shared document. Members may be added or deleted from groups and associated access rights are automatically assigned or removed.

The ALL-IN-1 Distributed Sharing Option allows the sharing of all file cabinets on a single system and across systems.

Set Mail User

The ALL-IN-1 Set Mail User option lets a user perform many mail operations on behalf of another user, such as a secretary might do for a manager. The Grant Mail Access Option lets users specify others who may process their mail. In this case, the mail header displays both users' names when a mail message is sent.

Addressing

Addressing enhancements include the following user interface improvements:

- The ability to create a distribution list using the list of addresses from a received message header
- Improved nickname support for longer nicknames

■ Easier manipulation of the long mail addresses common on messages sent or received via gateways to other mail systems

■ The ability to create a nickname from the sender of a message

Message Router

The Message Router OpenVMS gateway allows ALL-IN-1 users to exchange messages with users of the OpenVMS Personal Mail Utility. In addition, X.400 (1984) P2 attributes are supported between ALL-IN-1 IOS and the X.400 (1988)-compliant MAILbus400 message transfer agent.

Printing

The printing facility lets the user specify where to print the document, the number of copies, and the output format. The output can be directed to any of the printers available to the ALL-IN-1 IOS user, to another document, or to the user's terminal.

User Profile

The user e-mail directory contains an individual profile of every ALL-IN-1 user. A user can access and modify certain parts of the user profile, including:

Full name	Workweek and working hours
Address	Mail notification
Telephone number	Editing style
Department	Read receipt handling
Title	

The system manager maintains information such as user privileges, the default directory, and form library access.

International Support

ALL-IN-1 is designed to support multilingual operations, subject to availability of the specific translation. ALL-IN-1 can be deployed on a multilingual network; it can also support multilingual operations on a single system. However, all language kits, including U.S. English, must be the same version.

Time Management

Time Management makes it possible to maintain a personal calendar of events and tasks. An Events at a Glance screen gives the user an editable picture of the day's events, which makes it easy to make changes. Likewise, the Tasks at a Glance screen presents an editable picture of the tasks for the day. These pictures are combined on an Events and Tasks screen where other Time Management operations can be performed.

Group Conferencing

An ALL-IN-1 IOS user interface has been applied to Digital's DEC Notes application to provide the Group Conferencing feature. This feature lets users communicate on a many-to-many basis while taking advantage of ALL-IN-1's ease of use. Application developers can benefit from DEC Notes' integration at the API level in developing business and professional-specific applications that use this Conferencing capability.

Compound Document Support

ALL-IN-1 supports compound documents using the DEC Compound Document Architecture. This capability enables users to construct messages that not only contain simple text but graphic images, spreadsheet files, and word processing documents.

Online Training

VT users can take advantage of the online training facilities to learn how to use the system. The training guides users through sets of interactive screens and explains how the various features of ALL-IN-1 work.

Printing

Printer-level checking ensures that the print destination supports the document's final form. If the destination does not support the final form, then the document is converted to a format that is supported, if possible. Otherwise, the user is advised to print to a different printer.

System Management

The System Management utilities are used to perform routine maintenance of ALL-IN-1, such as scheduling and rescheduling of housekeeping procedures, management of printers, direct maintenance of network and DDS profiles, metering, and quota management of the shared area. A system manager has the option of creating a non-OpenVMS privileged system administrator to manage the following: user accounts (create, delete, move, rename); document archiving; system distribution lists; scheduling of housekeeping tasks; monitoring of ALL-IN-1 exception and information reports; and initiation of ALL-IN-1 shutdown.

ALL-IN-1 supplies a metering facility that gives system managers the ability to collect resource usage within ALL-IN-1 sessions for predefined events.

ALL-IN-1 TeamLinks Connection V2.0

This server enables TeamLinks users to access the file cabinet drawers, folders, documents, messaging service, and distribution lists. VT users familiar with ALL-IN-1 require minimum retraining to use this feature.

Users running Windows and Macintosh desktops can use their accompanying personal productivity applications to transfer documents of any type to the ALL-IN-1 file cabinet for secure storage. They can also use existing distribution lists to send these documents via the messaging service.

TeamLinks Mail is designed to let the user choose either DEC MAILworks or ALL-IN-1 Integrated Office System (IOS) as the Mail Service. TeamLinks Mail message exchange with VMSmail and ALL-IN-1 IOS is accomplished by means of the VAX Message Router. Either the DEC MAILworks Server or the ALL-IN-1 Mail service is layered on top of Digital's VAX Message Router to interact with all the other MAILbus gateways and complementary products.

TeamLinks incoming messages can be forwarded and/or answered. Answers can be directed to the sender only, or to all recipients of the message. Users can request confirmation that the message has been received and/or read. A message can be assigned one of three priority levels for delivery: Express, First Class, or Second Class. Alternatively, users can specify that the message delivery be deferred to a time set by the sender.

TeamLinks Mail for Windows provides customizable X.400 mail service and distributed file cabinet functionality along with document and graphical conversion services. As with other mail products, TeamLinks Mail performs the traditional tasks of creating, sending, and storing mail. It is a Windows GUI application that lets users exchange messages and attached files with other PC, video terminal, Macintosh, and workstation users across an enterprise WAN. TeamLinks Mail for Windows implements the 1984 CCITT X.400 P2 international messaging standard. Compliance with these standards extends the PC user's reach into the corporate enterprise WAN.

TeamLinks Mail for Windows permits remote and local file cabinets; transport over DECnet, TCP/IP, or non-DECnet; asynchronous launching of applications from within the mail application; and multiple attachments. Attachments can be binary files, such as spreadsheets and word processing documents.

Message Functions

Users can create, read, forward, reply, delete, and send messages using the menu commands. Auto-answer (IOS mail only), auto-forward, and read receipts are also supported. Message attachments may include another message or file. An appropriate gateway can be invoked to access and transfer this type of mail message. Messages can be addressed to one or more VMSmail, ULTRIX mail, and/or X.400 users worldwide.

The Windows and Macintosh TeamLinks products permit multiple Create and Read Message windows. They also allow you to print, edit, and read a message you created and its attachments prior to sending it.

Attachments

Files stored in the file cabinet can be sent as attachments to a mail message. Also, DOS files, such as word processing files and spreadsheets, can be mailed as attachments. To display these attachments, TeamLinks Mail either launches the appropriate document viewer or the associated DOS application.

Addressing

During message creation, users can simply type in one or more recipients. They also can select mail addresses from Digital's MAILbus Distributed Directory Service (DDS), a Personal Address Book (PAB), a distribution list, or the ALL-IN-1 subscriber list. All four services can be used in combination during message creation to automatically validate or look up a mail address.

DDS, available as part of the VAX Message Router, links TeamLinks Mail with other mail agents and gateways that share its directory services.

Filing Services

The distributed file cabinet model presents the user with one logical file cabinet structure. It provides storage on the DEC MAILworks server and/or on the ALL-IN-1 IOS server. It also allows storage locally in drawers, which contain folders that hold mail messages and/or documents.

The converters for TeamLinks reside on the PC. TeamLinks-supported conversion services include Mastersoft's document converters and Halcyon Software's graphics converters.

Standards Support

TeamLinks for Windows supports Microsoft Windows' Dynamic Data Exchange (DDE) and Dynamic Link Library (DLL) standards. This feature allows other Windows applications, such as Microsoft Word for Windows and Microsoft Excel for Windows, to access the TeamLinks X.400 e-mail service, file cabinet filing system, and document conversion services.

Notifications

Specifying delivery and receipt notification when you send a message generates a delivery report when the message is delivered or received.

You can set indicators to give the recipient information about the nature of the message, including importance (urgent, medium, low) and sensitivity (not restricted, personal, private, company confidential). These message classes are defined in the X.400 P2 e-mail industry standards.

Platforms, Operating Systems, and Network Specifications

Platforms: Windows and Macintosh

Operating systems: OpenVMS, Macintosh, and Windows

Networks: DECnet protocol stack and TeamLinks Mail are supported by these TCP/IP
                     stacks:
                        3COM 3+Open TCP
                        Beame & Whiteside TCP/IP
                        Digital PATHWORKS for DOS (TCP/IP)
                        FTP PC/TCP
                        HP ARPA Services for DOS
                        Microsoft LAN Manager TCP/IP
                        Novell LAN WorkPlace for DOS
                        Sun PC-NFS TCP/IP
                        Ungermann-Bass Net/One TCP/IP
                        Wollongong PathWay Access for DOS

Mail transports: Message Router, MAILbus, and MAILbus400

Message User Agent

Three message user agents exist in this e-mail environment: terminal-based ALL-IN-1,
Windows-based TeamLinks, and Macintosh-based TeamLinks. As shown in Figure 6-9, the
TeamLinks graphical user interface is menu-driven with toolbars and icons. The interface
leads the user through the process of easily and efficiently creating a message.

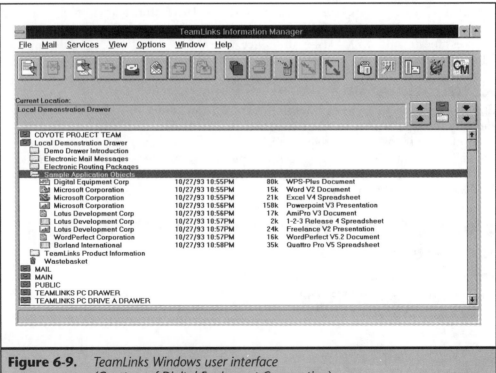

Figure 6-9.   *TeamLinks Windows user interface
                   (Courtesy of Digital Equipment Corporation)*

Product Directory Structure

The directory subsystem provides access to any directories available to the user and maintenance tools for ongoing administration. The user can view a directory, maintained by the system manager, that contains telephone numbers and addresses. A directory of all ALL-IN-1 IOS users can also be accessed. Facilities are provided for the user to maintain and view a personal telephone directory, nicknames, and distribution lists.

With the installation of Distributed Directory Service, ALL-IN-1 users can perform mail directory searches on defined criteria, such as a name, department, location, or organization.

Distributed Directory Services (DDS)

Most MAILbus gateways use Distributed Directory Services (DDS) to perform address translation for message originators and/or recipients in other vendors' messaging networks or in external networks. DDS provides utilities that allow replication and automatic updating of directory information across the Message Router nodes in a MAILbus network.

DDS optimally supports a population of up to 20,000 subscribers and/or 30 Message Router nodes.

Addressing Structure

Native addressing in the ALL-IN-1 environment is simple and similar to Internet addressing. It comprises three parts: the end user, the applications mailbox, and the node. An example of a typical ALL-IN-1 address is

John_Smith@A1@Node1

In this case, John_Smith is the end ALL-IN-1 user, A1 is the ALL-IN-1 application mailbox name, and Node1 is the name of the system on which the user's mailbox resides.

The ALL-IN-1 system also supports X.400 and SMTP Internet addressing.

Message Transport System

The message transport system in ALL-IN-1 is called Message Router. Message Router is similar to X.400 1984 in architecture and functionality. See Chapter 12, "X.400 Interpersonal Messaging," for a description of X.400.

Message Router is used in the ALL-IN-1 products. The newer X.400 1988 functionality is implemented with MAILbus 400.

Message Router

VAX Message Router (MR) is the core component of Digital Equipment Corporation's MAILbus Message Transfer System. MR is a layered software product that resides on OpenVMS VAX systems and provides three services to MAILbus messaging networks:

- Message Transfer Service
- Directory Service
- Management Service

Message Router and MAILbus are primarily used by interpersonal messaging (e-mail) applications. They can also be used by other applications, such as Electronic Data Interchange software.

Both the VAX Message Router VMSmail gateway (MRGATE) and the VAX Message Router are available as separate products. MRGATE provides a connection between MAILbus and the OpenVMS Mail Utility. VMSmail is the e-mail system packaged with the DEC operating system.

The MR Management Service helps with routine management of Message Router. It also monitors the mail network for error and exception conditions. Error and exception reports may be delivered to the MR network manager as events occur and/or on a scheduled basis. Management Action Procedures recommend corrective procedures for specific conditions. For complex networks, it is advisable to use the MAILbus Monitor, available from Digital's customer service, to assist in the management of the messaging network. The Monitor enables a system manager to know at a glance whether messages are reaching connected e-mail systems within the prespecified parameters.

Message Transfer Service

VAX Message Router provides a reliable store-and-forward transfer service for messages through a DECnet network. These messages may contain any information needing transport, including text, data, and arbitrary files. The message structure closely resembles what is described in the CCITT X.400 recommendations for message handling systems. However, messages use the encoding method described in the National Institute of Standards and Technologies (NIST) specification for the message format for computer based messaging systems, as shown in Figure 6-10.

In addition to Message Router, both the originator and the recipient need user agents, such as DEC MAILworks and ALL-IN-1 Integrated Office System, to create or read a message. MAILbus gateways connecting to other vendors' messaging systems, or to public messaging networks can also act as sources or destinations for messages. Organizations may develop their own applications using the VAX Message Router Programmer's Kit to send and receive messages through Message Router.

Message Router provides full support for delivery notification services and transmission of service and status messages. Network managers may establish routing schemes for messages by relying on default DECnet configurations, destination routing using route tables, or area routing for large complex networks involving interconnected hub nodes for each area.

Message Router contains the MAILbus Distributed Directory Service (DDS). DDS is a single, logical directory of subscribers and other MAILbus network information that can be distributed across a number of MR systems.

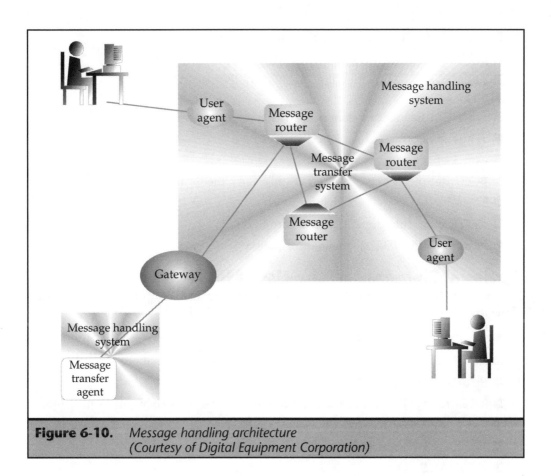

Figure 6-10.   *Message handling architecture*
(Courtesy of Digital Equipment Corporation)

Message Structure

The structure of the message router message is similar to X.400 1984, which is discussed in detail in Chapter 12, "X.400 Interpersonal Messaging." The following list itemizes the fields required in the envelope and heading, some of which are DEC-specific:

- *MESSAGE_ID*   MTA type identifier string; uniquely identifies the message across all platforms
- *UACONTID*   Content identifier
- *HOPCOUNT*   Number of times this message has been forwarded
- *CONTENTYPES*   Form of the message contents
- *PRECEDENCE*   The priority of the message: Low, Normal, or High
- *PDATE*   Posted date of the message

- *DEFERRED*   The earliest date the message may be delivered
- *ENCODEDTYPES*   The body part content types in the message
- *PERMSGFLG*   The action the Message Router should take when it receives the message
- *SENDER*   Sender of the message (this it not necessarily the From: field)
- *TO*   List of primary recipients
- *CC*   List of secondary recipients
- *BCC*   Recipients of which the TO and CC recipients are not aware
- *TRACE*   Intradomain trace information
- *INTTRACE*   Interdomain trace information
- *PERDOMAININFO*   Domain information

Optional Elements in the User Message Envelope

In addition to the required fields, the ALL-IN-1 message can also contain one or more of the following optional elements, as shown in Figure 6-11:

- *A1_TYPE*   File cabinet type
- *AL_FORMAT*   Document handling information for destination
- *SUBJECT*   Subject
- *MSGCLASS*   Purpose of message: Sensitive, Personal, Private, or Confidential
- *FROM NAME*   The account from which the message came
- *USERID*   User identification
- *ORGNAME*   Organization
- *FREEFORM*   String name (freeform)
- *TELEPHONE*   Telephone number
- *CREATED*   Date created
- *PDATE*   Posted date
- *PRECEDENCE*   Priority: Urgent, Medium, Low
- *TO NAME*   Message recipients

Name Format

The names in the Message Router have many components that are similar to those in the X.400 standards, to which it is closely related. Some of these components are required; others are optional. The USERID field is required in originator envelope name fields; the USERID and PERRECFLG fields are required in the recipient name fields. The fields that compose the address strings in the message router allow you to specify either ALL-IN-1 or X.400 addressing. These fields include:

- *USERID* User ALL-IN-1 identifier
- *ROUTE* Routing information
- *X121ADDRESS* Numeric address form
- *TERMINALID* Qualifier of the X121ADDRESS field
- *ORGNAME* Name of the organization

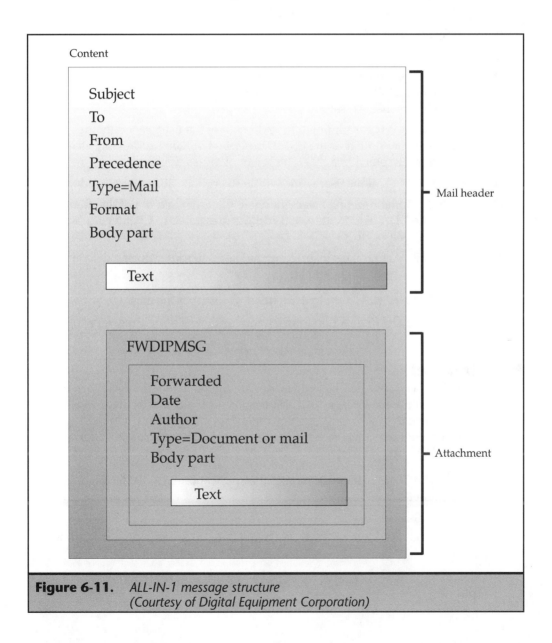

Figure 6-11. *ALL-IN-1 message structure*
(Courtesy of Digital Equipment Corporation)

- *ORGUNIT* Name of the organization unit
- *SURNAME* User's last name
- *GIVENNAME* User's given name
- *INITIALS* Users initials
- *GENERATION* Generation
- *TELEPHONE* User's telephone number
- *LOCATION* User's location
- *COUNTRY* User's country
- *ADMNAME* Commercial administrative domain of the user, generally a VAN; similar to the ADMD in the X.400 address string (described in Chapter 12, "X.400 Interpersonal Messaging")
- *PRMDNAME* Private administrative domain of the user, generally the company's name; similar to the PRMD described in X.400 addressing string (described in Chapter 12, "X.400 Interpersonal Messaging")
- *UNIQUEUAID* Unique user name within the private administrative domain.
- *DDNAME* Domain-defined element name that is passed out of the domain; similar to the DDA field in the X.400 address (described in Chapter 12, "X.400 Interpersonal Messaging")
- *PERRECFLG* Action flags instructing the message router during delivery
- *INTDDNAME* Internal-defined name
- *ORIG_INTENDRECIP* Original intended recipient of the message
- *REDIRECTINFO* Redirection information needed to route a message to another user

Heading Fields

The fields in the User Message Content Header are the same ones that are part of the message heading for X.400 Interpersonal Messaging. These components are specific to the type of message; in other words, an electronic data interchange (EDI) message might have different fields than an e-mail message. See Chapters 11 and 12 for a description of SMTP and X.400 fields.

- *APP_MESSAGE_ID* Identification of the application; in X.400 terminology, the IPM identifier
- *SUBJECT* Subject
- *INREPLYTO* The APP_MESSAGE_ID of the message to which this message is a reply
- *OBSOLETES* The APP_MESSAGE_ID of the message that this message replaces

- *REFERENCE*   The APP_MESSAGE_ID of the message to which this message refers
- *MSGCLASS*   Purpose of this message
- *PRECEDENCE*   Importance of this message
- *AUTOFORWARD*   Indicator that the message is an auto-forward body part
- *SENSITIVITY*   Sensitivity of the message as indicated by three values: Personal, Private, and Company Confidential
- *PDATE*   Posted date of the message
- *EDATE*   Expiration date of the message
- *REPLYBY*   Date by which a reply is needed
- *DATE*   Creation date of the message
- *TO*   Primary recipient
- *CC*   Copy recipient
- *BCC*   Blind copy recipients
- *REPLYTOUSER*   Users who require a reply
- *FROM*   User who originated the message
- *AUTHOR*   User who created the message
- *A1_TYPE*   File cabinet type
- *A1_FORWARD*   Message may not be forwarded
- *A1_FUNCTION*   Function to be executed upon receipt of message
- *A1_FORMAT*   Handling to be used on this message

Body Part Types

The Message Router message can have different types of body parts. These are the ones supported by the Digital Equipment Message Router system:

- *TEXT*   Text
- *FORWARDED*   Forwarded message content type
- *FWDIPMSG*   Forwarded message of IPM type
- *G3FAX*   Group Three fax
- *TIF1*   Text Interchange Format 1
- *TIF0*   Text Interchange Format 0
- *VOICE*   Voice
- *TELEX*   Telex
- *TELETEXT*   Teletext
- *VIDEOTEX*   Videotex

- *WPSPLUS*   Word processing
- *DX*   DEC Document Exchange document type
- *RMSFILE*   OpenVMS file types, such as relative, indexed, sequential
- *DDIF*   Digital Equipment Corporation Document Interchange Format
- *ODIF*   Office Document Interchange Format from the Office Document Architecture (ODA) standard
- *SFD*   Simple Formatable Document
- *VOICENOTIF*   Voice mail notification
- *IA5TEXT*   Text using the international ASCII standard
- *DECBODY1 through DECBODY 40*   DEC registered body part
- *VENDOR1 through VENDOR16*   DEC registered non-DEC body parts

Message Store

The file cabinet is a hierarchical set of folders and files where messages and other documents can be stored. Several fields are implemented in the Message Router to better support the filing cabinet.

File Cabinet Fields

- *TYPE*   File cabinet type
- *FORWARD*   Not forwardable
- *FUNCTION*   ALL-IN-1 function to perform when the message is received
- *FORMAT*   Type of format handling to apply to the message

Gateways Supported

These gateways are directly supported by DEC; others are available from third parties:

ALL-IN-1 Core Services V1.0 for OpenVMS VAX V3.0
ALL-IN-1 IOS for OpenVMS VAX V3.0
MAILbus LinkWorks for DEC OSF/1 (an office solution that can be tailored to meet the requirements of a particular customer)
MAILbus ULTRIX Mail Connection V1.1B
MAILbus VAX Message Router/P gateway V1.3 (IBM's PROFS and OfficeVision/VM systems)
MAILbus VAX Message Router/S gateway V1.3 (IBM's SNADS systems)
MAILbus VAX Message Router X.400 gateway V2.3
OpenServer400 cc:Mail[R] gateway to X.400 (a gateway between MAILbus400 MTA and cc:Mail)

OpenServer400 Microsoft Mail for PC Networks gateway to X.400 (a gateway between MAILbus400 MTA and Microsoft Mail for PC Networks)
OpenServer400 NetWare MHS gateway to X.400 (a gateway between MAILbus400 MTA and NetWare MHS e-mail system)
VAX Message Router VMSmail gateway

Fischer International Emc^2/TAO

Fischer International's strategic goal is to provide Emc^2/TAO for any computing configuration without compromising the benefits achieved from its client-server architecture and centralized, synchronized database.

Fischer International offers ways to effectively engineer the combination of e-mail and workflow with EDI, digital signatures, electronic document authorization (EDA), and authentication technologies. Why is this unique? Because forward-looking companies around the world are transacting "paperless" business today via electronic commerce—and it is the way every surviving company will be doing business tomorrow.

Fischer International will achieve its goal using its major products (Emc^2/TAO, Watchdog PC Data Security, WorkFlow.2000, and WorkFlow.2020) and its ongoing strategic alliance partner products, including MULTINET (Quadron Software International's EDI product) and SmartDisk (SmartDisk Security Corporation's smart card technology). Refer to Figure 6-12 for an example of Emc^2/TAO's architecture.

In strategic alliance with Motorola, Inc., Fischer International has redefined the uses of e-mail and paging technology. Emc^2/TAO messages and calendars can now be transmitted to alphanumeric pagers through any paging service in the world.

Emc^2/TAO, currently compliant with the 1984 version of the X.400 standard, will introduce 1988 compliance in 1994. X.500 is also currently under development.

Product Contact Information

Vendor:	Fischer International Systems Corporation
Address:	4073 Mercantile Avenue
	Naples, FL 33942
Phone:	(800) 237-4510; (813) 643-1500
Fax:	(813) 643-3772

Product and Version: Emc^2/TAO 3.4

Emc^2/TAO, whose architecture is client-server by nature, executes on DOS, CICS, CMS, IDMS, IMS, TSO, and VTAM with a common user client interface. It also executes on Windows and OS/2. The database that holds the user profiles, in-box, and file cabinets is located on the server. Emc^2/TAO's single database eliminates directory synchronization issues. The database is bit image-compatible across all Emc^2/TAO products and can hold a minimum of 100,000 users per server, with 10,000 concurrent users.

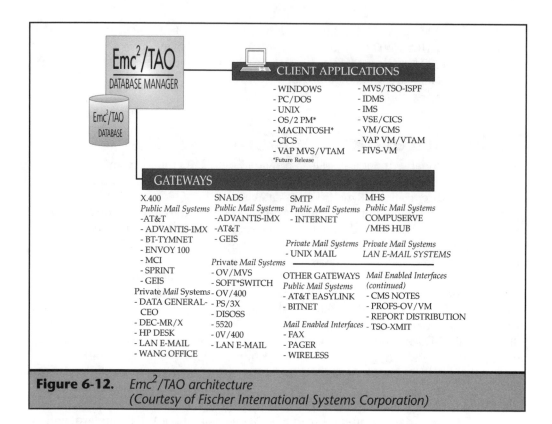

Figure 6-12. *Emc²/TAO architecture*
(Courtesy of Fischer International Systems Corporation)

Calendar

Emc²/TAO graphical calendaring features are an efficient way to organize your daily schedule and keep track of appointments, commitments, and special events. Calendaring allows you to schedule a full 24-hour day. After data on daily, weekly and monthly routines is entered, Emc²/TAO automatically generates a schedule. You can modify this schedule by adding, deleting, or rescheduling appointments, meetings, and private time. You can arrange meetings by simply choosing attendees or a public/private mailing list and indicating the date, time, and length of the meeting. Emc²/TAO then automatically notifies those invited. The product also facilitates resource scheduling, such as reserving A/V equipment and conference rooms. It also supports things-to-do and project management lists, with descriptors such as item title, due date, responsible person, and priorities.

In addition to terminal interfaces, Emc²/TAO supports two PC interfaces: DOS and Microsoft Windows.

The Personal Emc² remote interface allows users to compose, read, and file messages offline from the Emc²/TAO mainframe. When the PC connects to the host in-box and out-box, messages are automatically exchanged between the host and the PC. The PC interface has its own editor and directory address book.

Forms

Emc^2/TAO supports the design and transfer of "fill-in-the-blank" electronic forms. Any number of users can supply information on the same Emc^2/TAO form. Emc^2/TAO forms may be displayed, sent, forwarded from user to user, printed, filed, and reused. Forms are prepared like any other letter and any user can design them. Electronic forms may include highlighted and/or protected fields and entries; they can be of any length. Dates, times, and mail IDs can be entered automatically.

Conferences and Bulletin Boards

Conferences and bulletin boards may be used for sharing information and ideas, brainstorming, and decision-making. Members contribute by sending a letter addressed to the conference or bulletin board. The resulting information, arranged by date and contributor, becomes a permanent record of the process. Conferences and bulletin boards may be used as electronic file cabinets that support projects and workgroups. Emc^2/TAO allows an unlimited number of conferences and bulletin boards. Conferences may be organized by topic; bulletin boards may be arranged by name or department. Each has an administrator who may restrict access by class, department, or name.

Multilanguage Support

Emc^2/TAO supports American English, Brazilian Portuguese, British English, French, German, and Italian screen prompts and text. Spanish and Finnish will be available during 1994.

Platforms, Operating Systems, and Network Specifications

Platforms:	DOS and Windows
Operating systems:	Amdahl's UNIX (UTS)
	CICS on MVS and VSE
	CMS on VM
	IBM's UNIX (AIX) (future release)
	IDMS on MVS
	IMS on MVS
	OS/2
	OS/400
	PC-DOS and MS-DOS TSO/ISPF on MVS
	TSO on MVS
	VTAM on MVS and VSE
	Windows
Networks:	Banyan IP, IPX, and NetBIOS
Mail transports:	Proprietary database structure

Message User Agent

Emc^2/TAO has terminal- and PC-based message user agents. The Emc^2/TAO message user agent interface is shown in Figure 6-13. It includes the ability to view the calendar menu, contents of the user's in-box, the user's schedule for the day, and a view of the user's To Do list. The Windows version offers a toolbar and icons to facilitate this.

Product Directory Structure

The Emc^2/TAO product uses one central directory that services all the system's clients. The directory is composed of records. Each record in the Emc^2/TAO directory is a user profile and includes these fields:

- *Mail ID*   Unique ID of the user, bulletin board, or conference; it is 20 characters in length and uses this character set: _ / # $ - ' + * % ¦ <>

- *Duplicate Mail ID*   Unique combination of eight alphanumeric characters that allows the ID to be assigned to multiple users; the same as an alias

- *Department*   User's department; up to 20 characters long

- *Title*   User's title; up to 26 characters long

- *Password*   User's password; up to 16 characters long

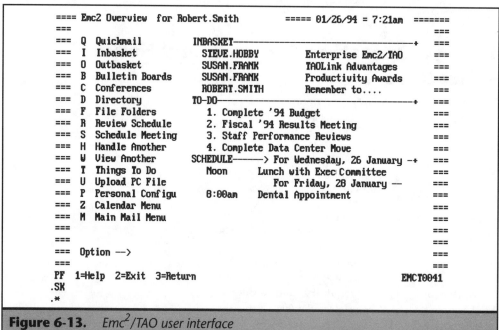

```
==== Emc2 Overview  for Robert.Smith        ===== 01/26/94 = 7:21am  =======
===                                                                      ===
=== Q  Quickmail          INBASKET--------------------------------+      ===
=== I  Inbasket             STEVE.HOBBY          Enterprise Emc2/TAO      ===
=== O  Outbasket            SUSAN.FRANK          TAOLink Advantages       ===
=== B  Bulletin Boards      SUSAN.FRANK          Productivity Awards      ===
=== C  Conferences          ROBERT.SMITH         Remember to....          ===
=== D  Directory          TO-DO-----------------------------------+      ===
=== F  File Folders          1. Complete '94 Budget                       ===
=== R  Review Schedule       2. Fiscal '94 Results Meeting                ===
=== S  Schedule Meeting      3. Staff Performance Reviews                 ===
=== H  Handle Another        4. Complete Data Center Move                 ===
=== W  View Another       SCHEDULE-------> For Wednesday, 26 January -+   ===
=== T  Things To Do          Noon      Lunch with Exec Committee          ===
=== U  Upload PC File                       For Friday, 28 January —      ===
=== P  Personal Configu      8:00am    Dental Appointment                 ===
=== Z  Calendar Menu                                                      ===
=== M  Main Mail Menu                                                     ===
===                                                                      ===
===                                                                      ===
=== Option —>                                                            ===
===                                                                      ===
PF  1=Help  2=Exit  3=Return                              EMCT0041
.SK
.*
```

Figure 6-13.   *Emc^2/TAO user interface*

- *Display Overview Menu as Main Menu* Sets which menu to show as first screen
- *Statistics & Accounting Read Authority* Specifies if user may read account data
- *Statistics & Accounting Reset Authority* Specifies if an API can sign on to this ID and reset accounting data
- *Suppress Display of System News* Permits user to prohibit display of system news items
- *Suppress Display of Department and Class News* Permits user to prohibit display of department and class news items
- *Auto-News at Logon* This feature enables a user to receive news items when first logging onto the system
- *Able to Send Un-Restricted Mail* Users may delete messages, edit messages, forward messages, export messages, print letters, and distribute outside of their own organization
- *Compress 3270 Data Streams* Data compression feature that the user can select

Addressing Structure

Addressing consists of Mail IDs and aliases. The Mail ID is a twenty-character address that is usually of the form:

First_name.Last_name

An example of a Mail ID is:

Rick.Smith

An alias is an eight character string that is unique to the server database directory, such as:

RSMITH

The alias might match the SNADS DEN eight-character string used to address users.
    Freeform addressing is also supported, such as addressing a message to a fax machine's telephone number. The format of such an address is

Rick.Smith@214-555-1212.FAX

Addresses can also be public or private mailing lists, bulletin boards, and conferences.

Message Transport System

The message transport system varies depending on the system in use, as indicated in the list of supported gateways.

Message Structure

The message structure of the Emc^2 product is proprietary. It contains the same TO, FROM, BCC, CC, Date, and Subject fields seen in all e-mail products.

Interface Issues

Functionality with other e-mail systems across gateways depends on the format, the gateway's protocol, and the characteristics of other systems.

Gateways Supported

Emc^2/TAO supports gateways to the following products:

ALL-IN-1	MCI Mail
AT&T EasyLink	Motorola Site Connect Server—
AT&T Mail	wireless pager
BITNET	NJE Node Support
BT Tymnet	Novell MHS
cc:Mail	OfficeVision/400
CMS Notes	OfficeVision/MVS
The Coordinator	OfficeVision/VM
Data General CEO	PROFS
DaVinci eMail	QuickMail
DEC Message Router	SMTP
DISOSS	SNADS (full support for notes,
EASYGATE	messages and DIA documents)
Envoy 100	Telemail
Fax	VMSmail
Higgins	WANG Office
HP Open Mail/HP Open DeskManager	

HP Open DeskManager

Hewlett Packard historically has offered the DeskManager product, which runs on the HP3000 series machines and uses proprietary message formats and transport protocols.

In addition to HP Open DeskManager, HP also presents HP OpenMail on the UNIX platform. This application provides a native implementation of X.400 1988, UNIX SMTP, and X.500 1988 (with proprietary 1993-like extensions). Both HP Open DeskManager and HP OpenMail offer complete integration of user agents, such as cc:Mail and Microsoft Mail, and gateways that ensure the two mail servers interoperate seamlessly on the same LAN.

Product Contact Information

Vendor: Hewlett Packard
Address: Nine Mile Ride
             Wokingham
             Berks RG11 3LL
             UK
Phone: 44-344-763100

Product and Version: HP Open DeskManager

HP Open DeskManager, the next release of HP DeskManager, is the electronic information distribution and messaging solution for HP 3000 MPE/iX systems and servers. Although significant enhancements have been made, HP Open DeskManager remains highly compatible with earlier versions of the product.

Both the HP Open DeskManager and HP Open DeskManager PLUS provide an open client-server messaging backbone that supports industry-standard protocols, open APIs, and a choice of client interfaces. Clients can be connected to the HP Open DeskManager backbone and directly communicate with each other. The PC clients communicate with the HP server using a new API specifically optimized for this purpose: the DUAL-DeskManager User Agent Layer. This direct client-server connection does not require a gateway when accessing HP Open DeskManager from PC LAN mail clients, such as cc:Mail or Microsoft Mail. The LAN mail post office is completely replaced, with HP Open DeskManager providing all necessary server functions.

The HP Open DeskManager C.00 release supports the following clients:

- HP clients

    HP AdvanceMail
    HP NewWave Mail
    Terminal/Terminal Emulation

- Third-party clients

    Lotus cc:Mail for Windows
    Microsoft Mail for Windows

The future direction of the product includes:

- Serial client support
- Support of additional third-party clients
- Apple Macintosh
- Microsoft Windows
- Novell NetWare support
- XWindow (MOTIF)

To third-party clients, such as Microsoft Mail and Lotus cc:Mail, HP provides specific software drivers called Windows Dynamic Linked Libraries (DLLs) that replace those supplied with the standard product. The DLLs redirect the client communications from the LAN post office to the HP 3000 server, while preserving the functionality of the client. Individual end users notice little difference in client functionality.

New Functionality

The following features are available with this release:

- Lotus cc:Mail client support
- Microsoft Mail client support
- Native mode port
- HP DeskNote
- UI enhancements
- AUTOPRINT
- Response time logging
- HP Spell (only available in HP Open DeskManager PLUS)
- Update DL server (only available in HP Open DeskManager PLUS)

Printing

AUTOPRINT is for users who receive "hard copy" e-mail. These users do not have a mailbox on the system but can receive messages in printed form. All incoming messages are printed automatically by the system printer for distribution; the original messages are deleted once they are printed.

SUPERPRINT enables users to have their messages print automatically, but still be able to send or reply to messages electronically.

A new server automatically updates personal and private distribution lists throughout the network.

File Attachments

Users can easily attach PC files, forms, spreadsheets, or any other form of electronic information—even video or audio—for transmission across the network. HP Open DeskManager is already capable of conveniently and reliably transporting multimedia and compound documents.

Connectivity

HP Open DeskManager offers wide connectivity to other systems through Hewlett-Packard and third-party gateways. These include connections to ALL-IN-1, fax, Internet/Sendmail, OfficeVision, PROFS, Telex, Wang, and X.400 as well as most leading public e-mail providers.

Filing Cabinet

HP Open DeskManager includes a sophisticated filing cabinet that supports user-defined folders and a wastebasket.

Forms Routing

HP Open DeskManager provides advanced workflow automation functionality with forms routing and signatures.

Platforms, Operating Systems, and Network Specifications

Platforms:	HP 3000 Series, HP 9000 Series
Operating systems:	MPE/iX Version 4.0 and higher
Networks:	HP 3000 NS for HP 3000 to HP 3000 connectivity, TCP/IP, modem for User Agent (client support)
Mail transports:	Proprietary, file transfer-based protocol used for all message transfer within the HP Desk network; provides native, peer-to-peer transport among HP 3000s acting as nodes within the e-mail network

Message User Agent

A number of message user agents (clients) are available for use with HP Desk, ranging from character-based terminal access to client-server GUIs, such as cc:Mail and Microsoft Mail.

Product Directory Structure

The Open DeskManager's directory comprises several datasets and index files, as shown in Figure 6-14. The main fields that tie the datasets together are listed in Table 6-1.

Field Name	Dataset Name	Field Description
BUILDING	GLOBAL-USER	Additional user info
CLASS-NAME	RESOURCE-INDEX	Name of resource class
CLASS-NAME	RESOURCE	Class of resource
CTRY-NAME	COUNTRY	Country code
CTRY-NAME	LOCATION	Country code
CTRY-TITLE	COUNTRY	Full name of country
FOREIGN-ADDRESS	FOREIGN-INDEX	
FOREIGN-ADDRESS	FOREIGN-ALIAS	Foreign address
FOREIGN-BLOCK	FOREIGN-ALIAS	Dataset block number
FOREIGN-COUNT	FOREIGN-ALIAS	Number of characters used
GATEWAY	GATEWAY-INDEX	
GATEWAY	FOREIGN-ALIAS	Gateway name
LOCKWORD	GLOBAL-USER	
LOCN-FLAGS	LOCATION	
LOCN-LANGUAGE	LOCATION	
LOCN-NAME	LOCATION-INDEX	Name of this location
LOCN-NAME	NODE-XREF	Location code
LOCN-NAME	LOCATION	Location code
LOCN-OFFSET	LOCATION	
LOCN-TITLE	LOCATION	Full name of location
MAIL-STOP	GLOBAL-USER	User information
NAME-AND-NODE	NAME-NODE-INDEX	Combined name/node
NAME-AND-NODE	FOREIGN-ALIAS	Name/mail node
NAME-AND-NODE	RESOURCE	Name/mail node
NAME-PROBE	NAME-INDEX	For name search
NAME-PROBE	GLOBAL-USER	For remove user name search
NODE-NAME	NODE	Name of mail node
NODE-NAME	GLOBAL-USER	
NODE-SEQUENCE	NODE-XREF	Sublocation code
NODE-TITLE	NODE	Full name of node
NUMERIC-ATTRIBS	RESOURCE	Up to four attributes
NUMERIC-NAMES	RESOURCE-INDEX	
SIMPLE-ATTRIBUTES	RESOURCE	Up to four attributes
SIMPLE-NAMES	RESOURCE-INDEX	
STRING-ATTRIBS	RESOURCE	Up to four attributes
STRING-NAMES	RESOURCE-INDEX	
TELEPHONE	GLOBAL-USER	Additional user info
USER-NAME	GLOBAL-USER	User name

Table 6-1.   *The Main Fields That Tie the Open DeskManager's Datasets Together*

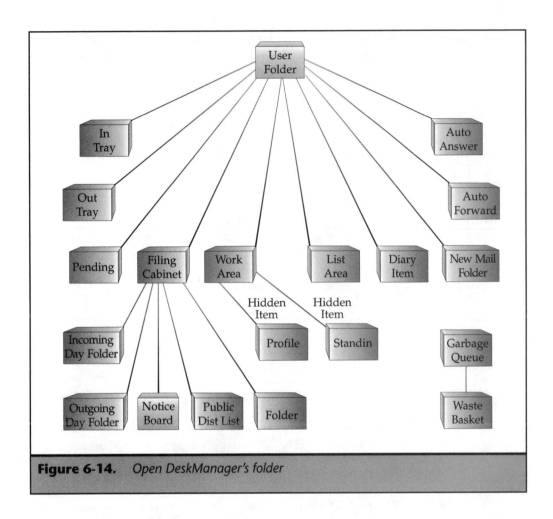

Figure 6-14. *Open DeskManager's folder*

The actual data fields used in the datasets listed in Table 6-1 make up the basic directory system information:

Dataset Name	Field Name	Field Length	Field Description
COUNTRY	CTRY-NAME	8 characters	Country code
	CTRY-TITLE	36 characters	Full name of country
LOCATION-INDEX	LOCN-NAME	6 characters	Name of location
NODE	NODE-NAME	8 characters	Name of mail node (location name left-justified and sublocation justified)
	NODE-TITLE	28 characters	Full mail node name
NAME-INDEX	NAME-PROBE	4 characters	Name for name search

Dataset Name	Field Name	Field Length	Field Description
GLOBAL-USER	NODE-NAME	8 characters	Mail node names
	NAME-PROBE	4 characters	Name probes
	USER-NAME	36 characters	User name
	LOCKWORD	2 digits	
	TELEPHONE	20 characters	User-configured
	MAIL-STOP	8 characters	User-configured
	BUILDING	8 characters	User-configured
	FILLER	44 characters	
NODE-XREF	LOCN-NAME	6 characters	Location code
	NODE-SEQUENCE	2 characters	Sublocation code
LOCATION	LOCN-NAME	6 characters	Location code, sorted
	LOCN-TITLE	28 characters	Full name of location
	CTRY-NAME	8 characters	Country code for search
	LOCN-FLAGS	1 digit	
	LOCN-OFFSET	2 digits	
	LOCN-LANGUAGE	2 characters	
NAME-NODE-INDEX	NAME-AND-NODE	44 characters	Concatenated name and mail code
GATEWAY-INDEX	GATEWAY	8 characters	Gateway name
FOREIGN-INDEX	FOREIGN-ADDRESS	128 characters	Foreign mail system address
FOREIGN-ALIAS	NAME-AND-NODE	44 characters	HPDesk name/mail node
	GATEWAY	8 characters	Gateway for this name
	FOREIGN-ADDRESS	128 characters	Address detail blocks 1-4
	FOREIGN-BLOCK	1 digit	Identifies address blocks 1-4
	FOREIGN-COUNT	1 digit	Character count for block
RESOURCE-INDEX	CLASS-NAME	16 characters	Name of resource class
	NUMERIC-NAME	4 x 16 characters	Name of numeric fields
	SIMPLE-NAMES	4 x 16 characters	Name of simple fields
	STRING-NAMES	4 x 16 characters	Name of string fields
RESOURCE	NAME-AND-NODE	44 characters	Unique name and mail node string
	NUMERIC-ATTRIBS	4 x 1 digits fields	Up to four attributes
	SIMPLE-ATTRIBS	1 digit	Bit mask field
	STRING-ATTRIBS	4 x 16 character fields	Up to four attributes

Mechanism of Directory Update

Directory updates are performed locally, either interactively or via batch file. Both mechanisms have the ability to generate a delta, or change, file that can be applied either automatically or manually to the directory on any other node within the network.

Addressing Structure

Names use the form given name, middle name, surname, and can include up to 36 characters (all elements). This limited number of characters can be quite restrictive, especially in countries where long names are the norm. Native addresses contain two components: a six-character location and a two-character sublocation, for example, HPCORP/UK.

A person's full address uses the form:

Richard Madeley/hp1600/01

In addition, there is an optional Foreign-Address field which is freeform and 128 characters in length. This field is used to address messages to recipients on external services or foreign mail systems. For example, you would use this addressing structure to send a message to a user on CompuServe:

Rik Drummond/hp1600/ux (7655.1044@compuserve.com)

In this example, hp1600/ux is the address of the Internet gateway, which will extract the Foreign-Address field from the distribution list when it receives the message from HP Desk.

The location/sublocation structure provides a limited hierarchy that can support multiple departments within a single logical entity. A single HP 3000 may contain multiple locations—such as hp1600, hp1400, and so on—each of which may have multiple sublocations.

Conversely, a single location may have multiple sublocations, each on a different HP 3000. In this example, hp1600 is a location that has sublocations (00, 01, and ux), each on a separate machine.

With this hierarchic form, you can address messages to residents of hp1600 without knowing the sublocation. In such a case, one of the systems must be designated as the node-resolving system.

For example, suppose Rik Drummond and Richard Madeley are resident on hp1600/01 and hp1600/02, respectively. The sender of a message only needs to know that both are located at the hp1600 site. The node that resolves for hp1600 sorts out the sublocation.

There is one drawback to this scheme, however. Unlike full-form addressing, where you can have two users with identical names (such as John Smith/hp1600/01 and John Smith/hp1600/02) who are differentiated by sublocation, hierarchic addressing requires each name to be unique.

Message Transport System

The HP Open DeskManager server message transport occurs via file transfers of the MPE files. The initiating end user's process posts a message in the out-box where the MAILROOM process decides the NODE QUEUE to which to attach the message. Messages bound for local users are attached to the local node's message queue; those bound for remote mail nodes are attached to that mail node's mail queue. A process called TRANSPORT ROUTER decides whether it is an MPE-to-MPE type message or a message going to a foreign system, such as a gateway. Messages traveling between MPE systems use a pair of processes called MASTER TRUCK and SLAVE TRUCK, on the source and destination systems respectively. These processes intercommunicate with the normal MPE interprocess communications (IPC) facilities. The IPC are the normal UNIX interprocess communications sockets that use the socket number/IP addresses pair to identify each node. These messages are delivered to the mail queue of the destination mail node. The MAILROOM process scans local mail queues and delivers the message to the local user's in-box. A user may choose to send a message with one of four priorities: Express, Urgent, Normal, and Deferred. Those with higher priority are transmitted and delivered before those with lower priority.

The message transport system architecture is similar to the SMTP architecture discussed in Chapter 11, "SMTP E-mail Services."

The MAIL TRUCKs can move information between MPE Desk systems by using either the DS or the NS protocols.

DS Protocol

Distributed Services (DS) was the first generation networking architecture for the HP 3000. Basically, it allowed each system to communicate with its immediate (directly connected) neighbors, but had no concept of domains or intelligent network level routing.

NS Protocol

Network Services (NS) succeeded DS in HP 3000 networking. NS makes a network appear fully connected at the application level, regardless of underlying physical connection. HP Open DeskManager had previously solved the problem and therefore does not make much use of this protocol.

Hewlett Packard has two ways of moving messages between MPE and non-MPE systems: intrinsic gateways defined by user written applications such as the OpenMail Gateway, and Foreign Service Connection (FSC), where the message is placed in a file to be processed by a user-written application. The MASTER TRUCK is responsible for posting these messages to the appropriate gateway entities.

The systems supports six different kinds of acknowledgments:

- *Transmission*   Transmitted off node
- *Receipt*   Message on the destination node
- *Delivery*   Message delivered to addressee's in-box
- *Read*   Addressee has touched the message through deletion, printing, reading, or opening
- *Reply*   Reply command or auto-answer has occurred
- *Forms reply*   Forms routing status

The Delivery functions like a delivery report, whereas Read is like a read receipt. Note that these status acknowledgments reflect the originating message user's point of view and are not messages; in other words, they show the status based on the original distribution list of the message. The user may look at the status of a message without cluttering the in-box with status messages.

Message Structure

The message structure is an MPE file. Three different structures define the message and hold the message heading fields: the ITEM-HEADER, the ITEM-STRUCTURE, and the ITEM-CONTENT databases.

ITEM-HEADER

- *ITEM-NUMBER*   A numeral identifying this structure to the MPE system
- *ITEM-TYPE*   Value that indicates whether the message is simple or complex; values less than zero are simple, those greater than zero are complex, multipart messages
- *ITEM-SUBJECT*   Subject of this item
- *ORIGINAL-ITEM-NO*   The ITEM-NUMBERs assigned when the message is first created on the home system; similar to the IPM Identification string in X.400
- *MESSAGE-ACTION*   Action required for the message
- *ITEM-COUNT*   Size of the total message for a simple message; not used for complex, composite messages
- *TOTAL-ITEM-SIZE*   Total item size in disk sectors of the message

ITEM-STRUCTURE

- *FOLDER-ITEM-NO*   The ITEM-NUMBER of the father item
- *CONTENT-ITEM-NO*   The ITEM-NUMBER of the son item
- *ATTACH-FLAG*   Bitmap showing status: read, unread, and so on
- *ACCESS-MASK*   Access privileges for a shared item

ITEM-CONTENT

- *ITEM-NUMBER* The ITEM-NUMBER of the basic item
- *ITEM-BLOCK* The data of the basic item
- *ITEM-COUNT* The length in sectors of the ITEM-BLOCK information

Message Store

The appearance of the message store varies somewhat depending on the client. For example, cc:Mail and Microsoft Mail message stores look like they normally do even when they're on the HP system. The message store supports:

- *In Tray* For receiving and reading incoming mail
- *Out Tray* For preparing and sending messages
- *Pending Tray* For monitoring the progress of messages
- *Work Area* For composing and editing messages and other items
- *Filing Cabinet* For storing and retrieving information in user-defined folders
- *Calendar/Diary* For time management and reminders

Gateways Supported

HP directly supports gateways for the following e-mail systems:

- HP Open DeskManager DISOSS & SNADS gateway
- HP Open DeskManager PROFS gateway
- HP Open DeskManager X.400 gateway

In addition, many other firms offer gateways for HP Open DeskManager products.

IBM OfficeVision/VM

With an installed base of millions, OfficeVision/VM historically has been IBM's e-mail offering. OV/VM runs on IBM mainframes under the VM operating system. It uses proprietary message formats, transport protocols, and directory structures. In addition, IBM offers several enhancements, including:

- Address Book Synchronization/2
- Time and Place Connectivity/2, which ties together OfficeVision/MVS, OfficeVision/VM, PROFS, and Time and Place/2 LAN calendars
- Current-OfficeVision/VM Workgroup Program, a client-server PC graphical user interface

Product Contact Information

Vendor:	IBM
Address:	5 West Kirkwood Boulevard
	Roanoke, TX 76299-0001
Phone:	(817) 962-5052
Fax:	(817) 962-3464

Product and Version: IBM OfficeVision/VM V1.2

OfficeVision/VM is compatible with PROFS version 2, Release 2, Modification Level 2, and higher. Its architecture is shown in Figure 6-15.

The OfficeVision/VM e-mail service enables users to create and exchange notes and documents throughout a network. List processing lets users process in-basket, document log, and other log items quickly by specifying and completing multiple tasks simultaneously. The new XEDIT-based note editor provides a powerful interface for creating, viewing, sending, and replying to notes. Users can store notes, documents, and information on the host computer for easy access and retrieval. Electronic calendar functions let users share personal calendars. Clear menus,

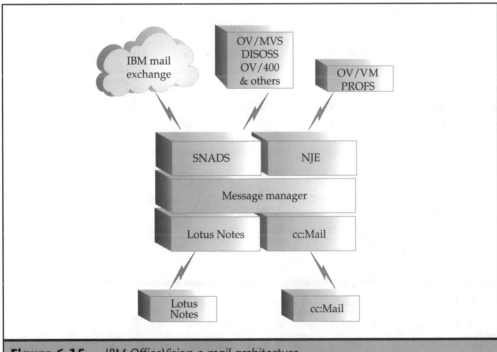

Figure 6-15.   *IBM OfficeVision e-mail architecture*
(Courtesy of International Business Machines Corporation)

messages, and help facilities make it easy for individuals to use OfficeVision/VM productively.

Calendar

Electronic calendaring services let you schedule your own appointments, group meetings, and conference rooms, all with a single operation. You can delete recurring meetings from your calendar. You can also review a calendar at the same time a meeting notice is placed on it so that conflicts may be resolved instantly. Meeting notices can be forwarded and directly added to the calendar. Other users may be authorized to access a calendar and add information. The new API allows applications to work directly with calendars. Calendars may be updated and viewed on other VM systems.

Time and Place Connectivity/2 version 1, release 2 is an OS/2 LAN group scheduling facility that can be used across hosts and LAN environments. A Time and Place/2 user can interact with the calendars of OV/MVS, OV/VM, and PROFS. This calendaring system is an actual database; it is not message-based like other products. As such, notifications and requests take place in real time.

Messages, Notes, and Documents

OfficeVision/VM supports three types of correspondence: messages, notes, and documents:

■ *Messages* are short, informal correspondence. They are not part of the person's incoming mail and no copies are saved.

■ *Notes* go to a person's incoming mail. They are more formal than messages, but less so than documents. A copy of the note is saved in the note log, unless this function is overridden by the user.

■ *Documents* are used for letters, memoranda, reports, and bulletins. They are formal correspondence. There are two types of draft documents: Document Composition Facility (draft) and Revisable-Form Text Draft (RFT-D). In addition, there are two types of final form documents: Document Composition Facility (final) and Revisable-Form Text Final (RFT-F).

Incoming notes and documents are marked with a type identifier; messages have no such identifier.

There are ten different identifiers for a note:

■ *Note*   Original note text

■ *Reply*   Reply to a previous note

■ *Forward*   Original note text with added text at top

■ *Meeting*   Text that can be added to a schedule

■ *Reserved*   Text that can be added to a conference room schedule

■ *Resend*   Note text received from someone else that may have been changed

- *Form* Notification of a form
- *Acklmnt* Acknowledgment of a message
- *Error* OV/VM error
- *Warning* Damaged note

Documents can be identified as any of the following:

- *Draft* DCF document that may be changed
- *Final* DCF document that may not be changed
- *RFT-D* RFT document that may be changed
- *RFT-F* RFT document that may not be changed
- *Paper* Hard copy document that was not created in OV/VM
- *Graphic* Graphical Data Display Manger (GDDM) soft copy file that can be printed or viewed
- *Other* Document that was not created in OV/VM
- *Deleted* Deleted document
- *Kanji* A document containing a graphic character set of Japanese symbols

These document and note identifiers are lost when the messages are transferred to other e-mail products that do not use them.

Filing Cabinet

Messages, calendars, documents, note logs, nickname files, document profiles, and directories may be filed in a filing cabinet. Every document is given a unique 12-digit number, called a CRON, when it is posted in the filing cabinet.

DOS Office Direct Connect

DOS Office Direct Connect gives PC users transparent access to host and PC services from the same menus. For cross-system communications, OfficeVision/VM can exchange mail with OfficeVision/2, OfficeVision/400, and OfficeVision/MVS, as well as with previous releases of OfficeVision and IBM PROFS.

Bulletin Board and Nicknames

Bulletin boards and nicknames are fully supported by the OV/VM system. Nicknames are eight-character sequences used to represent the *system_name.user_name*, or DGN.DEN sequence of eight characters followed by another eight characters.

Document Conversion and DisplayWrite/370

A DisplayWrite/370 document processor may be added to the OV/VM system. It allows conversion routines to interface so that documents may be shared among different word processors. See Chapter 16, "Document Conversion Options," for a discussion of document conversion technology.

Platforms, Operating Systems, and Network Specifications

Platforms:	IBM 370 architecture
Operating systems:	Conversational Monitor System (CMS) with VM/SP release 5 or 6, IBM Virtual Machine/System Product (VM/SP), VM/ESA, VM/HPO, and VM/XA release 2
Networks:	SNA
Mail transports:	IBM's RSCS

Message User Agent

The OV/VM product set supports two different message user agents:

- The historic OV/VM menu-like terminal interface used by PROFS since 1982.
- The IBM Personal Information Manager (PIM) Current-OV/VM Workgroup Program, which runs on Windows and OS/2 in the WIN-OS/2 environment.

Product Directory Structure

OfficeVision/VM systems typically use an IBM proprietary directory named CallUp. The structure and the directory synchronization technique are described in the following sections.

CallUp Directory

CallUp is a directory that is often used in VM systems. It need not have OV/VM running to operate; however, most OV/VM users use the CallUp directory to manage their e-mail addresses and other information.

Directory Synchronization

Directory synchronization occurs in a hierarchical manner, with one master CallUp directory on the network and many shadow directories. Any updates to a shadowed directory result in updates to the master directory, that take place either in real time mode or in a batch determined by the administrator. The master directory distributes the updates to the other shadowed directories to complete the update process.

Directory Record Format

The record format requires only three fields and is primarily selected by the user. The user name, a last modification time stamp, and a unique record number field are required. The other fields are all user-definable. The primary record length is 1,024 bytes, but you can extend a record by tying separate tables together using the unique record number field. Two types of fields exist in the directory: indexed and nonindexed. The maximum length of indexed and nonindexed fields are 45 and 255 bytes, respectively.

File Format

The OV/VM file format has three components: the directory definition record, the embedded index, and the actual data records.

Record Field Types

Record field types are generally user-definable. Most users choose to implement fields such as:

Full name	Phone
Organization	Fax
Reports to	E-mail address
Secretary	X.400 address

User-definable fields have no set field length. While this affords the system administrator some flexibility, the exchange of directory information with other systems is complex.

Addressing Structure

Addresses consist of an eight-character string that represents the distribution group name (DGN) followed by another eight-character string that indicates the distribution element name (DEN). This format is often called DGN.DEN and is used across the OfficeVision product line. DGN.DEN and PROFS Nodeid/Userid use the same addressing format, but with different names.

An example of a DGN.DEN address is

SYSTEM1A:MARKGOOD

Message Transport System

Messages are transferred between systems using Spool File Communications. The Remote Spooling Communications Subsystem (RSCS) is used for electronic message transfer, while Transparent Service Access Facility (TSAF) and APPC/VM VTAM Support (AVS) support intersystem calendar exchanges.

A mail message consists of ZIP packets. ZIP packets are 80 byte records that look like the 1403 card punch printer format. After the message is built, it is submitted to a remote distribution manager.

Distribution Manager

The message file is submitted to the virtual punch transport facility without a file header. The virtual punch is redirected to the RSCS local server, which is the intersystem MTA facility. The message file contains the destination node and user address in the DGN.DEN address format. The following example, in which CP stands for control program, shows sets of commands that would transmit the ZIP formatted message to another system using RSCS:

```
CP TAG DEVICE PUN'    target_node target_mailman;   Submit to virtual punch with
                                                    target_node and target_mailman
                                                    Tags

'CP SPOOL PUN TO'    local_rscs;                    Redirect the Virtual punch to
                                                    the local RSCS server

'PUN' zip_fn zip_ft zip_fr ' (NOH;                  Punch the File named zip_fn,
                                                    with file place on disk zip_ft
                                                    and file format zip_fr

'CP SPOOL PUN TO #';

'CP TAG DEVICE PUN';
```

Message Structure

Three kinds of message protocols are defined for the OV/VM system:

- E-mail message (ZIP5)
- Positive delivery acknowledgment (ZIPA)
- Error and nondelivery message (ZIPE)

Only the basic message is explained in detail here, but the other two formats generally have the same structure. A more detailed description of the ZIP5 packet, the electronic message format, appears in the following section.

ZIP5 packet is an electronic mail message, note, or document that is exchanged between PROFS systems. It includes fields that define the envelope, heading, and body, just like any other messaging protocol.

ZIPA is the delivery notification report packet format used to notify the sender that a positive delivery occurred. For example, if the ZIP5 packet had three addressees and the message was successfully transferred to each, three packets would be generated to report the deliveries to the sender.

ZIPE is the error and nondelivery notification report packet format. A packet is created for each error or nondelivery associated with a message.

These packets have an 80-character column format, much like the cards and printers of old. Consequently, the whole exchange appears as a series of 80-column units.

Envelope

The message envelope consists of several ZIP-type cards. The format of the ZIP5 e-mail message, not the ZIPA or the ZIPE formats, is described here:

- ■ ZIP5 is a single 80-character card that appears first and contains information such as the Document number, source DGN.DEN, and (usually) the PROFS filename.

- ■ ZIPN is a series of 0 to *n* 80-character cards that contains the DIA document profile information, such as Revisable Form or Final Form files.

- ■ ZIPU is a series of 0 to 256 80-character cards that contains the addresses of the destination in DGN.DEN format along with the location codes. ZIPU contains the To: and CC: addresses, but not the full names.

Heading

The ZIPP descriptor starts the message heading information. This artificial split occurs solely for the purpose of making the ZIP transmittal conform to the format of the other e-mail systems. The fields in this section are comparable to those found in other headings in other e-mail systems.

- ■ ZIPP is 0 to 5 cards in length and contains the text of the routing slips used with attached documents.

- ■ Record index is three cards long and contains information about the document, such as:

    Type of documents: acknowledgment, forward, re-sent, meeting, note, or reply
    Subject
    Author
    Attachment indicator
    Dates
    Whether it is a document or a note

- ■ CRON card is 1 to 281 cards in length and holds the author's and the addressee's full names, buildings, titles (such as Ms. or Mrs.), postscript titles (such as Ph.D.), and departments.

Body

The TEXT description contains the information being transmitted and is therefore the body of the message.

- TEXT is 1 to *n* cards in length; its format depends on the type of data being transported: text in 1403 card format or revisable form text.
- ZIPI is a one-card suffix record that ends the transmission of the message in the gateway.

Message Store

The OV/VM system has a full-function filing cabinet that supports the following commands: Sort, Find, Open, Delete, Retrieve Deleted Mail, Print, Search Note Contents, Split Folder into Two, Sort on Date, From, To, Subject, Type, Forward, Reply To, Resend, Range Search, Note Search Contents, Delegate, and Auto-Forward Mail.

Gateways Supported

Many other vendors support gateways to OV/VM. The IBM Mail LAN Gateway/2 supports the following products:

Advantis Network	OfficeVision/400
cc:Mail	OfficeVision/MVS
DISOSS	OfficeVision/VM via NJE
Emc^2/TAO	OV/MVS via DISOSS
Lotus Notes	PROFS via CMS
OfficePath/SNADS	PROFS via NJE

IBM OfficeVision/400

IBM's OfficeVision/400 e-mail system provides a consistent platform and interface for electronic communications. It also serves as a development platform for building companywide applications. OV/400 is a client-server solution that streamlines workflow and increases productivity within an organization. This office system features form processors, text editors, and graphics tools; it provides support for spreadsheets and faxes as well.

Product Contact Information

Vendor:	IBM
Address:	5 West Kirkwood Boulevard
	Roanoke, TX 76299-0001
Phone:	(817) 962-5052
Fax:	(817) 962-3464

Product and Version: OfficeVision/400 V2 R3

OfficeVision/400 is a workgroup platform that supplies e-mail, calendaring, and bulletin boards as part of its functionality. Fully integrated with the widely respected and reliable Application System/400, OV/400 provides a consistent platform and interfaces (both basic and intermediate) for complete communications, as well as an application integration platform for building enterprise solutions. It supports a client-server architecture and integration with other midrange, host, and LAN products. In addition, it enables IBM and non-IBM applications (such as spreadsheets, text editors, forms processors, and graphic products) to "plug into" the platform in an integrated manner.

E-mail

E-mail functionality includes the ability to send, copy, and receive documents, messages, and notes. Just like OfficeVision/VM, OV/400 supports three different types of messages: messages, notes, and documents.

- *Messages* are short, unformatted, unstructured correspondence.
- *Notes* use the standard e-mail format with Date, To, CC, and From fields, along with a text body. Notes are created with the messaging editor.
- *Documents* are created with a word processing editor.

E-mail supports different priorities (such as High, Normal, or Low) and different sensitivities (such as Personal or Confidential). Users can create and manage distribution lists and a personal address book. Users may also specify actions in messages such as: FYI, For your comment, For your signature, For your approval, Please handle, Please circulate, Please see me, and Please prepare a reply.

Calendaring

OV/400 allows you to schedule individual engagements, group appointments, and recurring events in a single operation. It supports automatic reminders, distribution of enterprise-wide meeting notices, and calendars for facilities, equipment, and other resources. Calendars may be printed or displayed by day, week, or month.

OV/400 supports several APIs that make it easier to integrate non-IBM products into the system. It offers two user interfaces: terminal mode and a client-server GUI interface called Current-OfficeVision/400.

Current-OfficeVision is a complete GUI application that runs under Windows or OS/2 and may be used on the OfficeVision 400, MVS, and VM platforms. It integrates the following services on the PC:

- *Calendaring*   For recording daily events and To Do lists
- *E-mail and document management*   For retrieval and storage of message and documents

- *Personal address book*   For recording names and addresses
- *Phone log and phone dialer*   For keeping notes about contacts and phone calls
- *Letter writer*   For using standard letter forms, including spell checking and contact management
- *Report generators*   For printing reports in a user-specified manner
- *Gantt charts*   For tracking projects and people
- *Outliner*   For organizing projects; this information may be transferred to personal calendars and projects lists
- *Graphics*   For viewing and saving clip art and images
- *Dynamic Data Exchange (DDE) support*   For integrating with other Windows applications
- *Import/Export*   For exchanging dBase, ASCII, and DIF files with other applications

Users may work offline and then post and receive information simply by signing on during the day.

Bulletin Board

Bulletin boards are supported among users on the same system. They are constructed in the same way as normal document storage folders.

Platforms, Operating Systems, and Network Specifications

Platforms:	AS/400 model 9404 and 9406
Operating systems:	OS/400
Networks:	SNA
Mail transports:	SNADS

Message User Agent

The Current-OfficeVision/400 graphical user interface is intuitive and easy to use. The application presents familiar desktop icons, a toolbar, and pull-down menus. The icons not only span the top of the user's screen but down the left side, providing quick access to many frequently used e-mail features.

Product Directory Structure

Directory information is contained in a file named QGFEMPSP on each OV/400 system. The file contains these fields:

Employee Name (8 digits)
Employee Last Name (10 characters)
Employee Middle Initial (1 character)
Employee First Name (8 characters)
Department Name (10 characters)
Work Phone (12 characters)
Home Phone (12 characters)
Title (15 characters)
Sales Quota (10 digits)
Birthday (6 digits)
Comments (21 characters)
Building Number (3 characters)

Floor (2 digits)
Room (5 characters)
Manager's Name (15 characters)
Manager's Phone (12 characters)
Secretary's Name (15 characters)
Secretary's Phone (12 characters)
Street Address 1 (22 characters)
Street Address 2 (22 characters)
City (12 characters)
State (2 characters)
Zip (5 digits)
User ID (8 characters)

Addressing Structure

The addressing structure consists of two eight-character sequences. The first string is the system address, such as DALXXB30, and the second is the user identification, such as JMSMITH. These are the DGN.DEN address strings used in the SNADS environment. This user's full address would appear as:

DALXXB30:JMSMITH

Message Transport System

The OV/400 system uses the same SNADS transport as discussed in Chapter 9, "IBM SNADS and DIA."

Message Structure

The overall message structure uses the Document Interchange Architecture and SNA Distribution Services (SNADS) discussed in Chapter 9, "IBM SNADS and DIA."

Message Store

The message user agents, terminals, and Current-OV interface all support normal foldering and file storage functions.

Gateways Supported

OV/400 supports gateways to the following products:

Advantis Network	OfficeVision/400
cc:Mail	OfficeVision/MVS
DISOSS	OfficeVision/VM via NJE
Emc^2/TAO	OV/MVS via DISOSS
Lotus Notes	PROFS via CMS
OfficePath/SNADS	PROFS via NJE

Lotus cc:Mail

Lotus is committed to open standards. The Lotus Communications Architecture (LCA), as illustrated in Figure 6-16, features the Lotus Communications Server (LCS). LCS is a cross-platform, multiprotocol messaging service based on cc:Mail and Notes technology. This adds native SMTP/MIME and X.400 1988 messaging as well as X.500

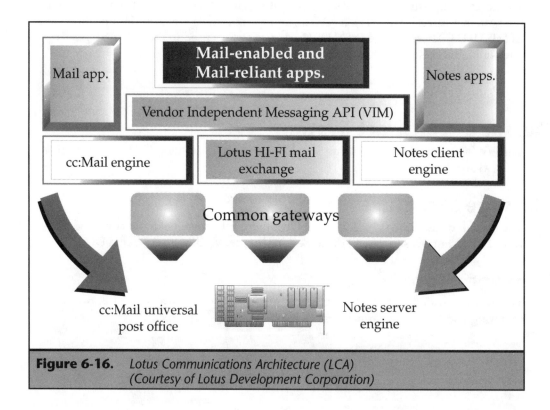

Figure 6-16. *Lotus Communications Architecture (LCA) (Courtesy of Lotus Development Corporation)*

directory support to both the cc:Mail and Notes product lines. Lotus will deliver LCS on DOS, Windows NT, OS/2, UNIX, and NetWare as a NetWare Loadable Module. The LCS products should be on the market by the first quarter of 1995.

Product Contact Information

Vendor:	Lotus Development Corporation
Address:	800 El Camino Real West
	Mountain View, CA 94040
Phone:	(800) 448-2500; (415) 961-8800
Fax:	(415) 961-0215
E-mail/BBS:	(415) 961-0401

Product and Version: Lotus cc:Mail

Lotus cc:Mail is a LAN-based e-mail system with more than five million mailboxes. It is one of the market leaders in the LAN e-mail environment. It supports MS-DOS, Windows, OS/2, UNIX, and Macintosh. The versions all support a common directory and messaging format and do not require a message gateway among them.

cc:Mail supports both small and large workgroups well. It also provides a variety of message routing configurations: hub, master/slave, and peer-to-peer.

cc:Mail supports several messaging APIs, including Vendor Independent Messaging (VIM), X.400 API Association's Common Messaging Calls (CMC), Microsoft's MAPI (API and SPI for Microsoft Mail and Chicago), and XAPIA-X/OPEN transport and directory interfaces.

cc:Mail consists of four distinct components: user agent, message transport, directory services, and mail engine. The design allows easy integration with other products, such as Banyan StreetTalk and Novell Global MHS directory services.

cc:Mail's Fault Tolerance feature provides sophisticated self-running diagnostics that pinpoint and resolve problems before they become visible to the user.

cc:Mail offers multiple levels of encryption on each post office to ensure security and privacy. All messages are encrypted during inter-post office transmittals. Administrators can easily relax or strengthen security. The product permits administrative control of minimum password length, password expiration, and the number of incorrect log in attempts allowed.

Comprehensive statistics on directories, user activity, disk space usage, and telephone connections are also available to the administrator.

cc:Mail supports addressing by name instead of a complex string. It also offers these features:

- TO, CC:, and BCC: addresses
- Return receipts when a user prints, reads, or deletes message
- Three priorities: Low, Normal, and Urgent

- Distribution lists
- Personal mailing lists
- Direct addressing of messages to folders

cc:Mail for MS-DOS

The MS-DOS Platform Pack contains both user and administrative software to set up and run on a DOS platform. DOS users may exchange messages and share files, text, and graphics with cc:Mail users on Windows, OS/2, UNIX, and Macintosh platforms.

cc:Mail for MS-DOS supports TSR, spell checking, keyword message retrieval, and highly automated management functions. It can use the Novell Bindery, Banyan StreetTalk, or cc:Mail Directory services. There is also a mobile version of this product.

cc:Mail features context-sensitive help and an interface that anticipates the user's most likely actions.

Through TSR and application integration, cc:Mail can launch most DOS applications from an associated file attachment or "on the fly" to create new files. cc:Mail users can also "take a snapshot" of a DOS application's screen and share it with other cc:Mail users as a graphic item.

You can swap cc:Mail out to 17K of memory and hot-key among mail and other applications. You can also receive incoming mail notification while using other applications.

cc:Mail for Windows

cc:Mail for Windows is a full-functioned system that supports the standard Windows features, such as pull-down menus, one-click short cuts, and help facilities that use a Hypertext approach. cc:Mail can be running in one window at the same time that other programs are executing. cc:Mail supports the following features:

- Cut and paste functions
- File attachments, which appear as applications icons in cc:Mail messages; you double-click an icon to launch the source application
- Viewing of files created in many popular applications from within cc:Mail, even if the applications are not installed on your system
- Notification of new mail
- Use of multiple Boolean operators, such as AND and OR

cc:Mail fully supports DDE and uses DDE macros to mail-enable other Windows applications, such as Microsoft Excel, Word, and others. Many of these macros are available on the commercial networks. You can also call cc:Mail from within other applications.

cc:Mail supports a full-function, built-in text editor with adjustable fonts and automatic word wrapping, much like the Windows Notepad.

cc:Mail for OS/2 Workplace Shell

cc:Mail for OS/2 Workplace Shell is an object-oriented e-mail environment that is tightly integrated with IBM OS/2 version 2.1. It takes full advantage of OS/2's drag-and-drop capabilities and object-oriented file management system. As a result, all cc:Mail functions, including in-boxes, message folders, bulletin boards, and directories, appear as individual Workplace Shell objects on the OS/2 desktop. Versatile Message Template Objects let you send mail to a preset mailing list directory from the desktop. The product is optimized for OS/2.

cc:Mail for UNIX

The cc:Mail UNIX Platform Pack includes a graphical user interface that supports cc:Mail- and SMTP-based systems. Users of SUN SPARCstations and 100% compatibles can create multimedia messages and file attachments under OPEN LOOK. The Platform Pack contains both UNIX- and DOS-based administrative software to support cc:Mail. It handles up to 200 bulletin boards and uses the UNIX Network File System (NFS) to store and retrieve files and messages.

cc:Mail enables communications across DOS, Windows, OS/2, and Macintosh platforms. It supports mail lists and in-box. You can set up cc:Mail to notify you by beeper, audio files, or another means when new mail arrives. Addressing uses the simple cc:Mail address format. cc:Mail permits complex searches for messages, bulletin boards, and file cabinets. You can perform searches based on read/unread status, author, date, priority, or file attachment.

cc:Mail for Macintosh

The Macintosh Platform Pack contains both user and administrative software to set up and run cc:Mail. Macintosh users may share messages with cc:Mail users on other networks running MS-DOS, Windows, OS/2, or UNIX.

Mobile Mail

cc:Mail supports mobile mail for MS-DOS, Windows, Macintosh, and HP 100LX Palmtop platforms.

Time Management

Lotus Organizer is a Windows-based product that uses the cc:Mail directory to view user calendars and schedule meetings. Attendees are notified by cc:Mail of the proposed time, location, and subject. The person may accept, decline, defer, or delegate the meeting.

Platforms, Operating Systems, and Network Specifications

Platforms: DOS, Windows, Unix, and Macintosh
Operating systems: DOS, Windows, OS/2, UNIX, and Macintosh

Networks:	cc:Mail supports these networks:
	AlisaShare
	AppleShare
	DEC Pathworks
	LAN Manager
	NetWare
	Paceshare
	TCP/IP for DOS from FTP software
Mail transports:	cc:Mail Router

Message User Agent

The cc:Mail Remote user interface is shown in Figure 6-17. The Windows version offers a rich graphical user interface. This intuitive interface uses SmartIcons, which provide one-click shortcuts to e-mail tasks. You can customize the icons to execute the tasks you perform most frequently. You can also launch other applications from within cc:Mail by simply clicking on the designated icon.

Product Directory Structure

The directory connects through Import/Export, the user agent, and the administrator functions. The Import/Export utility defines the directory content and uses a command-line interface for adding, deleting, and updating directory records.

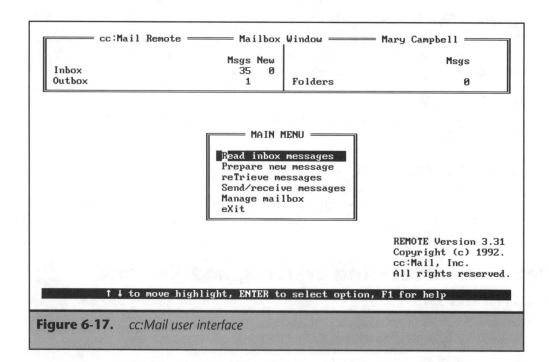

Figure 6-17.   *cc:Mail user interface*

Overall, the directory has five fields:

- *Name*   User name or alias (126 characters long)
- *Loc*   Name location with respect to this post office (1 character)
- *Last Checked In*   Date and time last checked in (18 characters)
- *Comments*   General comment field (126 characters long)
- *Address*   cc:Mail address field (126 characters long; 255 for a foreign address name)

Addressing Structure

The basic fields for cc:Mail addressing are based on the location of the recipient and post office relative to the sender's post office where the directory copy resides. For example, you would appear as a local user in your post office's directory, but as a remote user on a different post office's directory. cc:Mail does not have an address string, as such.

Many mail systems, such as ALL-IN-1, use a descriptive full name and the entire address for each user, as shown here:

Full name: Rick Smith
Address: Rick_Smith at A1 at CORP

In cc:Mail, users are referenced by their full names or an alias. Addresses are relative to the current server. The full name is part of a record. The address record contains the user's foreign or remote address and comments. Sending mail to a local user requires inputting the user's name, as in:

Jack Smith

If Jack Smith is on another post office, the address table would show that post office's name, and reference Jack as:

Jack Smith at PO7

If Jack Smith is on a Novell network, he would be referenced as Jack Smith and the address table would indicate that he could be found in:

MHSLINK
M:\MHS\MAIL\GWAY
CORP

Table 6-2 shows the address fields of different types of users.

Address Type	Address	Example
Local user	(blank)	
User at remote post office	Post office name	COPR
cc:Mail remote user	(blank)	
cc:Mail remote user	Telephone	232-3456
Alias	User name	Smith,Rick
Fax user at FAXPO	Fax post office	FAXPO
MHS user	MHS post office	MPOrsmith@CORP [full name] {Rick Smith}
PROFS user	PROFS post office	PROFSPO rsmith

Table 6-2. *cc:Mail Address Types*

An export of a normal cc:Mail directory record might look like this:

Name: Smith, Ted   Addr: HQ-1
Locn: L   Cmts: Manager of Finance

where the following is true:

- *Name* is the name of the user or an alias
- *Locn* is the address relative to the post office specified in the Addr: field. Locn: uses the Universal Format of an address in this field and can have several values:

    L   Local user at the post office specified in the Addr: field
    R   cc:Mail Remote user at a post office specified in the Addr: field
    A   Alias of a user specified in the Addr: field
    P   Post office
    DL   Delete user
    DR   Delete cc:Mail Remote user
    DA   Delete alias
    DP   Delete post office

- *Addr* is the address of the user relative to the post office
- *Cmts* is the Comments field

Mechanism of Directory Update

The directory is updated through either the cc:Mail Import/Export utility or the Automated Directory Exchange (ADE) utility. Several directory update topologies are supported by ADE:

- Master/slave
- Peer-to-peer
- Advanced

In the *master/slave* topology, one post office receives updates from all the slave post offices and then redistributes all the modifications back out to the slaves. In the *peer-to-peer* arrangement, each post office sends its updates to all the other peer post offices. In the *advanced* approach, the directory synchronization may be a mix of both techniques.

The Import/Export utility may also load and unload information using flat ASCII files. cc:Mail uses Import/Export commands such as:

IMPORT M:\CCMDIR @E:\DIR\INPUT DIRECTORY

where IMPORT is the basic command; M:\CCMDIR is the file where the updates will go; E:\DIR\INPUT is the file containing the additions, changes, and deletions; and DIRECTORY indicates that the data is directory, not message, data. The same general command format is used to export directory entries.

Changes made to a local directory by the administrator generate an e-mail message at the end of the session. This message is stored on a special bulletin board, named ##Directory Updates, on the post office, and can only be viewed by the administrator. This bulletin board also holds the directory propagation list of other ##Directory Updates bulletin boards on participating post offices. cc:Mail Router extracts the relevant updates before sending the message to the other participating directories. This process is called directory update send filtering.

The cc:Mail mail router also performs directory update receive filtering as it receives inbound directory update messages. In this case, only information that is pertinent to the local mailbox is permitted through the filter to the local directory update programs. If the directory updates are configured properly, this inbound filtering should not occur.

ADE, the cc:Mail directory synchronization utility, also supports remote/mobile users' directory updates. Recently, ADE was enhanced to provide wide-area mail management capabilities in several areas:

- Flexible Directory Propagation allows organizations to set up topologies that match their structure, such as boundaries that shield specific address information from outside organizations, (for example, suppliers or competitors), and domains that let the administrator assign limited rights. For example, a bulletin board manager cannot alter the company's worldwide routing.

■ Control can be centralized, decentralized, or mixed; in contrast, most LAN-based e-mail products allow only a peer-to-peer topology, which can become unwieldy in networks with many post offices. For example, in a peer-to-peer topology, adding one user to each of 30 post offices generates 900 messages; with ADE 2.0, only 60 messages are created. In addition, peer-to-peer does not allow central control, for example, importing an external database, such as a human resources list.

■ Flexible synchronization provides fully automatic capabilities that can be scheduled. This eliminates the need, for example, to manually export the directory and then import it into another post office. Scheduling allows the administrator to optimize network traffic and set processing at convenient times.

■ In addition to directories and bulletin board messages, administrators can propagate mailing lists and bulletin board titles, even to mobile users.

■ The ability to use monitoring programs and provisions for future wide area mail management, such as remote administration of the call list, which is the schedule of actions for cc:Mail's router; remote administration of the post office profile and routing tables; and a common format for status, alert, and error messages with the ability to route them to a single mailbox.

■ Enhanced user interface options are specified on screen, making it easier for novice users to perform tasks.

Message Transport System

The cc:Mail Router supports the transfer of mail between post offices on the same file server, on the same LAN, and on the same WAN. It also lets remote cc:Mail users exchange mail by using either single or multisession router software.

Two types of routers, or MTAs, are supported:

■ *Type 1* is for transfers between post offices on the same LAN. This type of e-mail transfer uses one cc:Mail Router, which copies messages from disk to disk using the underlying network disk-sharing software.

■ *Type 2* is for transfers between post offices not on the same LAN. This exchange requires two cooperating cc:Mail Routers, one at each end.

The cc:Mail Router executables are usually stored in a server directory called M:\CCROUTER. The router program, RTADMIN, uses a call list that specifies the post office, gateways, and networks to be contacted for inbound and outbound mail as well as the time and day of the week they should be contacted.

Like the ADE, the cc:Mail Router may be configured in three different ways: master/slave, peer-to-peer, and advanced. In peer-to-peer configuration, each post office server has an associated cc:Mail Router that contacts all the other post offices and moves mail directly between the source and destination post offices.

In the master/slave configuration, the cc:Mail Router is associated with a single post office. This post office collects and posts mail for others in its call list group by picking up the mail, holding it, and then moving it to the destination server.

The advanced configuration combines the master/slave and peer-to-peer arrangements.

The cc:Mail Router software may be executed as a background task on a post office if the post office supports multitasking. It can also run on a separate PC associated with a post office when multitasking is not supported. In either case, a cc:Mail Router is associated with a post office—it is not independent.

Message Structure

The actual structure of the cc:Mail message is proprietary and not known outside of Lotus. However, because Lotus has a well-defined utility for importing and exporting messages, information on the exact message structure isn't necessary. Messages may be imported and exported to flat files using the Import/Export utility, which is the same tool used to import and export directory information.

Heading Fields

Fourteen fields comprise the envelope and heading of a cc:Mail message:

- *Message:* Beginning of the message; may be followed by numbers that indicate the actual size of the message; the message ends when another Message: field or an end of file is encountered

- *From:* Author of message (may be up to 126 characters long, followed by a PO name of up to 126 characters)

- *Forwarded by:* Person who forwarded the message

- *Date:* Date and time of the message; uses the format *mm/dd/yy hh:mmxx* where *mm* is the month, *dd* is the day, *yy* is the year, *hh* is the hour, *mm* is minutes, and *xx* is either AM or PM

- *To:* Address of main recipient with one name per To: field; cc:Mail allows up to a total of 200 To:, cc: *To:, *cc:, and BCC: recipients

- *cc:* Copy recipients

- *BCC:* Blind copy recipients

- **To:* Nonreceived recipient or recipients of the message that are not on the cc:Mail system

- **cc:* Nonreceived copy recipients or copy recipients of the message that are not on the cc:Mail system

- *Atinbox:* Alternate folder for placement of the message

- *Priority:* Priority of message: Low, Normal, or Urgent

- *Subject:* Subject of the message (60 characters long)

- *Receipt requested:* Indicates that the sender wants a read receipt

- *Not delivered to:* Name of unknown recipient

Body Part Types

You can attach any kind of file to a cc:Mail message. Up to 20 attachments of unlimited size are supported per message.

cc:Mail messages do not have multiple body parts; instead, they have attachments. These attachments are defined in the message heading in the same way as those in the envelope and heading fields:

- *Contents:* Beginning of message content; may be followed by a number indicating the length of the contents. Each line has a maximum length of 80 characters and ends with a CR/LF. The field ends with reserved words: Message:, Text item:, File item:, Fax item:, Graphics item:, or end-of-file.

- *Text item:* Beginning of separate text item

- *File item:* Beginning of a file included in the message, such as D:\Data\Info.wk1; may be followed by the message size count

- *Include:* Beginning of an ASCII text file

- *Fax item:* Beginning of a fax item

- *Graphics item:* Marker for a graphics item

Message Store

cc:Mail supports the common file cabinet with foldering capability.

Gateways Supported

cc:Mail supports these gateways; other gateways are available from gateway vendors:

3COM Mail	DEC ALL-IN-1
AT&T EasyLink	DISOSS
Banyan Network Mail	HP Open DeskManager
cc:Mail EZLink	Lotus Notes
cc:Mail Link to 3COM Mail	MCI Mail
cc:Mail Link to MCI Mail	Novell MHS
cc:Mail Link to MHS	OfficeVision
cc:Mail Link to SMTP	PROFS
cc:Mail Link to Soft*Switch	QuickCom messaging network
cc:Mail MHSlink	SMTP
cc:Mail PROFSLink	Sprint Mail
cc:Mail Telelink	UUCP
cc:Mail UNIX/UUCP	Verimation Memo
DASnet	Wang VS Office
DEC VMSmail	

Lotus Notes

Due to its tight integration with Lotus 1-2-3, one of the world's most popular spreadsheet applications, Lotus Notes has become a widely used e-mail system in corporations today. Many workflow applications have been developed using Lotus Notes as the underlying e-mail transport system.

Much speculation has surrounded Lotus Notes since Lotus Development Corporation acquired cc:Mail. Analysts predicted that Lotus would shift its development dollars to cc:Mail and away from Notes. This has not occurred thus far and Lotus continues to allocate development funds to the Lotus Notes product.

Product Contact Information

Vendor:	Lotus Development Corporation
Address:	161 First Street
	Cambridge, MA 02142
Phone:	(617) 577-8500
Fax:	(617) 693-4663

Product and Version: Lotus Notes Version 3

Lotus Notes is a client-server platform for developing groupware applications. One of its subsystems is NotesMail. Lotus Notes improves the business performance of people working together by enhancing the quality of every day business processes, such as customer service, sales and account management, and product development. Because Lotus Notes unifies the critical technologies required to create these applications, it offers developers an efficient platform for building strategic business process applications.

Lotus Notes is a robust, secure development environment in which developers can quickly create cross-platform client-server applications that have an immediate impact on the efficiency of business processes. With these knowledge-sharing applications, multiple users can manage compound documents, communicate effectively, and take advantage of business process automation over geographically dispersed or even remote locations. Lotus Notes lets users access, track, and share organized, document-oriented information. This information can include multiple formats, such as text, images, video, and audio. It can also incorporate files from outside sources, such as other desktop applications.

The Lotus Notes workspace is a graphical user interface that runs under Windows, OS/2, UNIX, and Macintosh System 7. Lotus Notes provides a consistent user interface across each of the platforms, with such features as pull-down menus, scroll bars, pop-up and multimedia annotation, and context-sensitive help.

The Notes application supports three kinds of databases: shared, local, and mail. Database information is replicated and synchronized across servers as part of the default Notes functionality. The system manager schedules updates at predefined intervals. The replications ensure that all the copies of a database become identical over time. Replication is also supported between desktop clients and the server. This feature is critical for employees at remote locations who infrequently attach to the server. It ensures that the employee's client maintains an up-to-date copy of the server database.

Notes system management supports performance statistics, alerts, event reporting, and remote operation of servers. It conforms to the Internet Simple Network Management Protocol (SNMP) standards.

NotesMail lets you communicate quickly and easily with other NotesMail users. You can send and receive messages, ranging from single-line documents to complex, multipage, compound ones. You can even attach an entire Lotus Notes database file to a message. NotesMail notifies the user when new mail arrives.

Each user has a mail database that stores all messages that are created, sent, and received. In Notes, a database is a single file in which each record is stored as a document. Lotus Notes includes a pull-down menu that gives the user access to the help facility, which consists of a Notes database and supports context-sensitive help. Lotus Notes uses RSA public key encryption to authenticate users, provide access control, preserve confidentiality, and identify sources and destinations.

Lotus NotesMail integrates with cc:Mail using the Lotus Mail Exchange Facility, which supports directory synchronization, document conversion, and message exchange between both products. Notes can exchange mail with other applications that use the Novell MHS facilities, such as BeyondMail, DaVinci eMail, and The Coordinator.

Notes supports the VIM API from Lotus. (See Chapter 18, "Electronic Messaging APIs," for more information about VIM.) A user on a VIM-compliant mail system may disable NotesMail and use the other system, such as cc:Mail, if the cc:Mail user interface is more familiar to the user.

Notes naturally supports forms and workflow functionality; in fact, NotesMail is a special example of a workflow form.

Platforms, Operating Systems, and Network Specifications

Platforms:	Windows, OS/2, and Macintosh for clients
	Windows and OS/2 for servers
Operating systems:	Windows, OS/2, UNIX, and Macintosh System 7
Networks:	AppleTalk, LAN Manager, NetBIOS, OS/2 EE, PATHWORKS,
	SNA TCP/IP, SPX/IPX, VINES, and X.25
Mail transports:	Lotus Notes Mail Router

Message User Agent

Lotus Notes presents a robust graphical user interface complete with pull-down menus, icons, a toolbar, pop-up windows, and dialog boxes. Figure 6-18 shows the Lotus NotesMail user interface for the Windows platform.

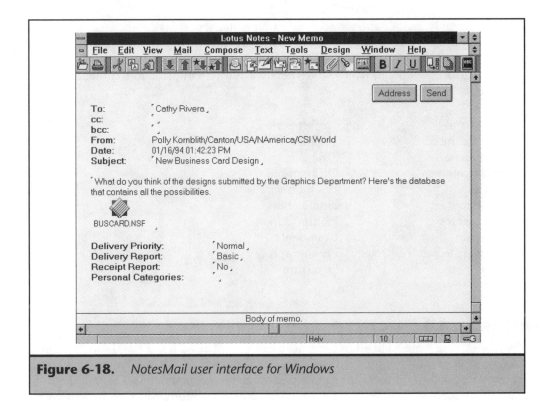

Figure 6-18.   *NotesMail user interface for Windows*

Product Directory Structure

The Notes directory is called the Name & Address Book database, which is a single file on the server. The file is composed of records, each one stored as a document.

All personal data is included in a document called a Person Document. Each Person Document stores a single e-mail user's name, address, and other pertinent information.

Name & Address Book

Directory updates employ the same replication facilities used to synchronize other Notes databases. The directory is called the public Name & Address Book. A full copy of the document is stored on each Notes server. In addition to user information, the public Name & Address Book includes the following information:

- How a user's mail should be delivered
- Groups and Access Control Lists (ACL)
- Intervals at which servers should communicate
- How to establish a connection between servers
- Schedules for replication of the Notes databases

A personal Name & Address Book exists on each client. This database contains private data, such as personal distribution lists, nicknames, and in the case of remote clients, phone numbers.

The Person Document, which stores personal address information, contains these fields:

First name	Department
Middle initial	Location
Last name	Title
Full name	Office phone
Owner	Fax number
Mail domain	Home phone
Mail server	Home address
Mail file	Comments
Forwarding address	Photograph
Short name	Signature
Company	

Addressing Structure

The underlying NotesMail addressing structure complies with the X.500 standard. In fact, it employs the distinguished name concept used in X.500. (See Chapter 8, "X.500 Directory Systems," for more information about distinguished names.) Even though a user may address another user by a name such as "Ben Smith," the system actually stores it in a distinguished name format similar to that used in the X.500 standards.

An example of a Notes user's name is

CN=Ben Smith/OU=Accounts Payable/OU=Finance/O=ABCTools/C=US

The system displays the name in the following manner:

Ben Smith/Accounts Payable/Finance/ABCTool/US

In the previous example, the abbreviations are the same as in the X.500 convention:

- *CN*   Common name or the user's given name and surname
- *OU*   Organization unit, such as Sales or Finance
- *O*   Organization, such as Acme Parts
- *C*   A two-character abbreviation for any country in the world

A name can have up to four OUs, with each successive one representing a higher level of the organization. In the previous example, Accounts Payable is an organization unit within the broader unit of Finance.

Message Transport System

Notes allows other mail systems, such as cc:Mail and other VIM e-mail applications, to replace NotesMail in Notes. In this case, the other mail systems use their own transport systems, not the Notes database update system, to transfer mail.

When a NotesMail user sends a message, a program called Mailer executes on the client. Mailer verifies the address and then places the message in the server's MAIL.BOX database. Each server participating in NotesMail has one MAIL.BOX. The Router program periodically checks MAIL.BOX for waiting mail. If the addressee is on the same server, it moves the message directly to the user's mail file, such as BSmith.NSF. If the user is on a different server in the same Notes network, Notes delivers the message to the other server's MAIL.BOX database. If the user is on a server in another Notes network or domain, or one that is not always connected but uses the same data communications protocol, the mail Router examines the Connection Document in the Name & Address Book to determine the next server that will connect to the destination. The Router then moves the message to the MAIL.BOX file on that server.

The mail Router is a store-and-forward mechanism that, depending on the information contained in the Name & Address Book database, moves the mail messages to intermediate servers to await final delivery.

The mail system supports on-demand mail delivery if a high priority message is sent. Also, the administrator may specify that mail be sent after a certain number of messages accumulate in the MAIL.BOX file.

Mail Delivery

Lotus Notes delivers mail differently depending on whether the sender and recipients are on the same or a different network.

Two servers are on the same network if they are connected with the same communications protocol and only have data communications routers or bridges in-between. If they have a gateway in-between, they are on different networks. See Chapter 13, "OSI, LAN, and Other Protocol Stacks," for a discussion of different communications protocols.

Users on the same network have mail delivered by the mail Router as soon as it encounters the message in the MAIL.BOX. Users on different networks generally wait for mail delivery until the next scheduled connection. These connections are usually database update connections.

Mail routing is a critical component in the Notes network. Routing enables the exchange of messages and maintains communications among databases in a Notes application. Figure 6-19 depicts an overview of the Notes routing scheme and the flow of mail from the originator to the recipient.

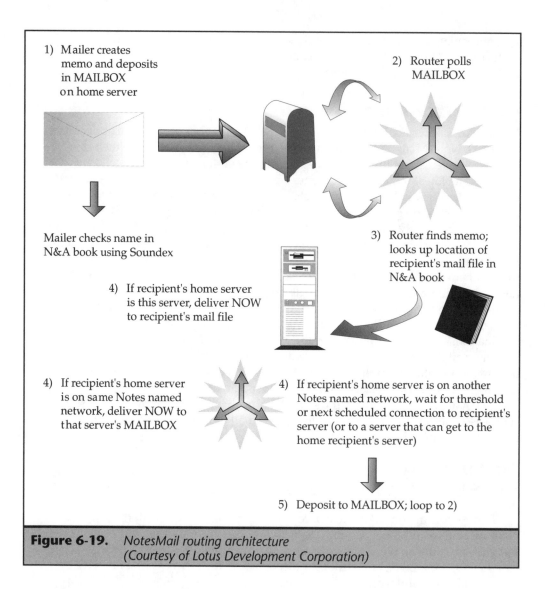

1) Mailer creates memo and deposits in MAILBOX on home server

2) Router polls MAILBOX

Mailer checks name in N&A book using Soundex

3) Router finds memo; looks up location of recipient's mail file in N&A book

4) If recipient's home server is this server, deliver NOW to recipient's mail file

4) If recipient's home server is on same Notes named network, deliver NOW to that server's MAILBOX

4) If recipient's home server is on another Notes named network, wait for threshold or next scheduled connection to recipient's server (or to a server that can get to the home recipient's server)

5) Deposit to MAILBOX; loop to 2)

Figure 6-19. *NotesMail routing architecture (Courtesy of Lotus Development Corporation)*

Message Structure

Lotus Notes documents are compound in nature. They are created and maintained through Notes forms. Forms allow the user to enter information in a structured format, just as you might enter it into a purchase order.

Five types of mail forms are standard:

- Memo
- Memo to Manager

■ Phone Message

■ Reply

■ Reply to All

Other types of forms may be created by the user, application developer, or administrator. This conceptual structure is similar to that used in the CE QuickMail message.

The message body is in Rich Text Format (RTF), which may contain text, enhanced text with color and fonts, tables, graphics (such as NSI Metafiles), scanned images, and voice objects. These various objects can be embedded using OLE; they can also be attached to a NotesMail message.

There are several default mail fields that are predefined in the NotesMail application. A message is created with the @MailSend Notes command (one of the Notes API-type commands), which contains parameters that match the fields described here. The only field that is absolutely necessary is the SendTo field; the others are optional.

The NotesMail fields are:

■ *SendTo*   List of To recipients

■ *CopyTo*   List of carbon copy (cc) recipients

■ *BlindCopyTo*   List of blind copy (BCC) recipients

■ *Subject*   Subject of the document

■ *Body*   Text of the message in Rich Text Format

■ *DeliveryPriority*   One of three priority levels: High, Normal, and Low

■ *DeliveryReport*   One of three reporting instructions: Basic (report generated only if delivery fails), Confirmed (report always generated), and No report (report never generated)

■ *ReturnReceipt*   Read receipt requested or No read receipt requested

■ *Encrypt*   Encrypt document or Don't encrypt document

■ *Sign*   Always sign document or Don't sign document

■ *MailOptions*   Mail memo

■ *SaveOptions*   Save to disk

■ *MailFormat*   Identifies whether it is a cc:Mail or NotesMail message and if it is Encapsulated (encapsulated in a Notes document), Text (document in text format), Both (both encapsulated and text format), or Notes Memo format

The mail router adds the following fields when the document is mailed:

■ *PostDate*   Time and date the document was mailed

- *RouteServers*   Sequence of servers through which the document passed on the way to its final destination
- *RouteTimes*   Total time the message was en route
- *FromCategories*   Contents of the Categories field of the attached Notes document, if present
- *DeliveredDate*   Date and time the document arrived in the database

Message Store

The three categories of Lotus Notes databases are shared, local, and mail.

Shared Databases

Shared databases reside on one or more Notes servers and can be accessed by many users. Security is used to grant and restrict database access to certain groups of users. These databases are synchronized across the network through a process called replication. Databases are replicated, or copied and distributed, to servers on the network.

Local Databases

Local databases reside on an individual Notes workstation. They are generally personal in nature and contain daily diaries or prototypes of new databases that are not yet ready to be shared. They can also be local replicas of databases stored on remote workstations so a user can access them without connecting by phone to the network.

Mail Databases

Mail databases usually reside on the Notes server, but only the user may access his or her own individual mail database. Users who access Notes through remote connections often store a replica of their mail database locally. NotesMail is useful for private conversations and small group communications outside of the public shared databases.

Gateways Supported

Lotus Notes directly supports the following gateways; other third-party vendors support additional gateways to Lotus Notes:

- Lotus Mail Exchange Facility for Notes Release 3 and cc:Mail for Windows 2.0
- MHS gateway

Microsoft Mail

Microsoft is the leading software vendor for personal computers. Microsoft Mail is Microsoft's LAN-based e-mail system that enables workgroups to communicate more efficiently. Microsoft Mail offers a robust user interface, connectivity to more than 50 other e-mail systems, and provides flexible tools for system administration. The product lends itself to deployment in geographically dispersed corporations, and offers a variety of remote access capabilities.

Product Contact Information

Vendor:	Microsoft Corporation
Address:	1 Microsoft Way
	Redmond, WA 98052
Phone:	(206) 936-8080
Fax:	(206) 93-MSFAX

Product and Version: Microsoft Mail V3.2

Microsoft Mail for PC networks v3.2 is a PC system supporting many enhanced e-mail features. The Windows version supports a simple text editor with spell checking, foldering, and subfoldering for message storage and retrieval, as well as several address lists for finding e-mail destinations. It allows administrators to define special templates for adding fields to the directory and create special message types specific to an application. Users may view faxes, create message templates, forward, read, save, create, attach files, and delete messages. Messages may be stored on the user's PC or on the server.

Calendar/Scheduling

Schedule+ is an add-on product to Microsoft Mail for PC networks that provides calendaring and scheduling. This software allows a user to schedule meetings and manage the calendar. It is easy for users to respond to meeting requests. Assistants or others can send and receive meeting notices on another user's behalf if they have the appropriate privileges. Additional administration of Schedule+ is minimal because of its tight integration with Microsoft Mail.

Bulletin Boards

TEAM Channels is an add-on product that provides online forums that can be accessed from Windows. Users can exchange files, questions, and answers. The features include private and public discussion groups, live message updates, drag-and-drop file attachments, filtering, and message threading.

Forms

An add-on product called Electronic Forms Designer, based on Visual Basic, allows users to design and send forms as e-mail messages. Special control buttons can be added to provide custom behavior. The forms can be automatically processed without human intervention. The form initiator will be notified as to who has the form in the form distribution list when the next recipient receives the form.

Another forms product by Delrina, called FormFlow, seamlessly ties installed applications and resources to existing databases and Microsoft Mail. FormFlow has flexible forms routing capability, which can establish rules and procedures for routing and coordinating all kinds of information. In addition, FormFlow can be integrated with Paradox, Clipper, ASCII, and SQL databases.

Remote User Agent

Mail Remote for Windows is a software program for remote access to a user's mail box. This product features the same user interface as the LAN-based Microsoft Mail for PC networks product. A remote user can read and create messages offline as well as work with the messages in the message store.

Workgroup

Workgroup Templates is a collection of templates based on products like Word and Visual Basic. The package also includes the source code for the templates.

Document Conversion

KEYview from Keyword provides users with instant formatted views and printing of over 35 leading word processing, spreadsheet, and image file formats. KEYpak from KEYWORD offers one of the most comprehensive document conversion facilities on the market. See Chapter 16, "Document Conversion Options," for more information on Keyword's products.

WinRules

WinRules from Beyond Incorporated allows you to automatically file messages in designated folders, auto-answer, and auto-forward messages to designated users.

Multimedia Mail

Use MailRoom from Simplify Development Corporation or Watermark Discovery Edition from Watermark Software, Inc., to improve office productivity; both allow embedding of image and fax data in the message, allow annotations to the images, and support OLE.

Platforms, Operating Systems, and Network Specifications

Platforms:	Intel processors
Operating systems:	OS/2 and MS-DOS/Windows
Networks:	LAN Manager, IBM LAN Server, and NetWare
Mail transports:	External, DEC Pathworks, Banyan, 3COM, Novell MHS, and Global MHS

Message User Agent

Microsoft Mail's end-user interface is graphical and intuitive, as shown in Figure 6-20, making it very easy to use to perform routine e-mail functions. The presentation screen contains a toolbar, pull-down menus, an informative in-box, and icons for shortcut access to e-mail facilities such as moving a message to a folder.

Product Directory Structure

Microsoft Mail supports several address lists for directory lookups. They are

■ *Global*   This directory contains the whole network; it may include users on other mail systems

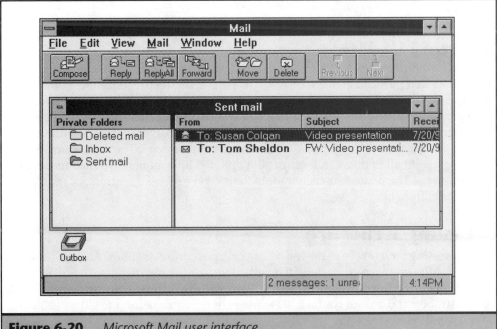

Figure 6-20.   *Microsoft Mail user interface*

- *Personal*   This is a user's personal address book
- *Post office*   This is a list of all users on one post office

Basic Directory Fields

The three directories each have six default fields. Others may be added on a site-by-site basis. The six default fields are

- *Alias*   Distribution lists and user names (30 characters long)
- *Name*   User name (30 characters long)
- *Address Type*   Type of application the message is for, such as Microsoft Mail
- *Mailbox*   Mailbox name (10 characters long)
- *Post office*   Post office name (10 characters long)
- *Network*   Network name (10 characters long)

Many sites create a site directory template for input of additional user information. The additional fields are often such optional items as the following:

Employee Identification	Zip
Phone	Country
Fax	Department
Street Address	Supervisor
City	Division
State	

Directory Update

A single post office in the organization is the directory server or *master directory*. Other post offices that participate in the directory synchronization process (a post office may or may not participate) are called *requesters*. At predetermined intervals, a network-wide program, called DISPATCH, starts the extracts on the requester post offices. The requester post offices extract directory data from the post office address list and mail it to the directory server in a specially formatted mail message. The master directory server processes all updates and returns the updated information in a mail message to the requesters. The requesters then update their server's global address lists with the information from other servers.

Addressing Structure

The addressing structure is of the form network, post office, mailbox. Each of these names may be up to ten characters in length.

The mailbox name is usually set to the user's server's logon name to simplify administration and reduce confusion. However, it need not be, and one user could have several addresses.

A mailbox name could be BenSmith or SmithBen. Post office and network names could be CorpHQ and ABCTool, respectively.

A Microsoft Mail address looks like the following to a user:

BenSmith/CorpHQ/ABCTool

where BenSmith is the mailbox name, CorpHQ is the post office name, and ABCTool is the Microsoft Mail network name.

Message Transport System

Microsoft Mail's message transport system uses a program called EXTERNAL to transmit messages between servers.

EXTERNAL

EXTERNAL is the program that runs on separate MS-DOS systems or on OS/2 multitasking systems. To move mail, EXTERNAL uses the underlying shared file system of the NetWare, Banyan, and LAN Manager products. EXTERNAL maps the server disk drive to the post office that Maildata resides on and then moves mail between the two systems. It moves mail by doing disk to disk copies. EXTERNAL periodically logs on to the post offices in its list and checks the post office's mailbags for new mail. After finding new mail, it moves it to the destination server. Sometimes mail messages are moved to an intermediary server. Other EXTERNALs will later move the mail message to the final destination.

VOLUME DIRECTORY STRUCTURE   The default directories that the Microsoft Mail files reside in are C:\MAILDATA for mail messages and distribution lists, (the post office) C:\MAILEXE for the server executables, and C:\MAILUSER for the client files.

The basic subdirectory structure of the Microsoft Mail post office, C:\MAILDATA, is composed of 19 subdirectories.

Subdirectory	Description
ATT	Contains message attachments for all users
GRP	Group (distribution lists and pointer files)
KEY	Index files that contain pointers to the header records in the mailbag files held in MBG (one key file per user)
MEM	Contains the member of the local post office global group that is used by EXTERNAL
TPL	Template files that are kept and correspond to INF files in INF
HLP	Help files
LOG	Log files and undelivered message files
MMF	Microsoft Mail File store that is in your in-box and is saved for windows users (one file for each user)
USR	User and group names of address of external users

FOLDERS	Folders, such as public and private folders
INF	Information files such as ADMIN.INF, the user address data file
MAI	In-box for everyone's text part of the message and addressees
NME	Name the files of the alias address list (one file per user)
XTN	Contains external post office information
GLB	Contains the Microsoft Mail system information, such as group names, user access information, gateways, networks, modem information, and queues for directory synchronization transactions
MBG	Mail headers that point to the mail text in the MAI directory
P1	Directory that contains temporary files that buffer inbound and outbound messages being moved by EXTERNAL

The volume directory structure supporting the e-mail message logically looks like the following:

- *C:\maildata\mbg*   Mail bag holding all inbound users, mail with one file per user. The file names are in hex code, such as 00000000.mbg, representing users and post offices. Users such as Dave, Sally, or PO1.

- *C:\maildata\p1*   Temporary heading records are kept here for outbound and inbound messages being moved by EXTERNAL, the Microsoft Mail MTA.

- *C:\maildat\mai*   Text is kept for the in-box messages in files in subdirectories from ma0-ma9 and maA-maF. The subdirectories contain the text for retrieved messages named by temporary hex file names, such as 000094D1.MA9.

- *C:\maildata\att*   Inbound and outbound attachments are stored in subdirectories from at0-at9 and atA-atF, by temporary hex file names, such as 00004456.ATT.

Message Structure

The message structure is defined through the MAPI routines in Microsoft Mail for PC Networks v3.2. The message is stored as a whole in the user's file folders. It is stored as separate components, headings, text body, and attachments in the in-box and out-box. See Chapter 18, "Electronic Messaging APIs," for a discussion of MAPI.

General Message Structure

MAPI calls are used to create a mail message. The message contains the following fields:

- *lpszSubject*   Pointer to subject string of up to 256 characters

- *lpszNoteText*   Pointer to message text string

- *lpszMessageType*   Pointer to message type string indicating type of applications the message supports; a null pointer indicates an interpersonal message (e-mail)

- *lpszDateReceived*   Date/ Time message is received in *YYYY/MM/DD HH:MM* twenty-four hour format

- *lpszConversationID*   Pointer to the coversation identification string
- *flFlags*   Bit field used as flags to indicate none to all of: Unread, Receipt Requested, and Sent
- *1pOriginator*   Pointer to MapiFileDesc structure that describes the originator of the message
- *nRecipCount*   Number of Recipients pointed to by the lpRecips field
- *lpRecips*   Pointer to array that describes the recipients of the message
- *nFileCount*   Number of attached files pointed to by the lpFiles field
- *lpFiles*   Pointer to the array of file attachment descriptor

Address Encoding

Addresses have the following form and are described by another MAPI routine— *MapiRecipDesc*. The fields of the routine are

- *ulRecipClass*   Type of recipient: Originator, To, Cc, or BCC
- *lpszName*   The name as displayed by the message system
- *lpszAddress*   Provider-specific message delivery data; need not be a Microsoft Mail address
- *ulEIDSize*   Size in bytes of data in lpEntryID
- *lpEntryID*   Microsoft Mail–specific binary representations of lpzsAddresses; not interpretable

Attachments

Attachments are appended to the message by the MAPI routine *MapiFileDesc*. It has the following fields:

- *flFlags*   Indicates the type of OLE objects referenced by the attachment
- *nPosition*   Position of the attachment in the message body. The attachment replaces the character at the specified position in the MapiMessage routine defined structure, described previously.
- *lpszPathName*   Pointer to the full path name of the file, for example, M:\data\month.dat
- *lpszFileName*   Pointer to the file seen by the recipient. lpszPathName could be a temporary file. If it is NULL, this one replaces it.
- *lpFileType*   Pointer description to type of attached file. Currently this is NULL.

Message Store

The Microsoft Mail post office is composed of common folders and subfolders. These may reside on the workstation or the server.

Interface Issues

Microsoft Mail uses characters in the address string that many systems do not support.

The network/post office/mailbox address string components each have a maximum length of ten characters. Exchanging information with other systems that do not have this address length, without an address converting switch or gateway in between, will cause problems.

Gateways Supported

Many third parties support gateways for Microsoft Mail. Below are Gateways offered directly by Microsoft:

Microsoft Mail gateway to IBM PROFS and OfficeVision/VM
Microsoft Mail gateway to MCI Mail
Microsoft Mail gateway to AT&T Mail
Microsoft Mail gateway to X.400
Microsoft Mail gateway to SNADS message services
Microsoft Mail gateway to SMTP
Microsoft Mail gateway to MHS
Microsoft Mail Connection for AppleTalk

Chapter Seven

Strategies for Connecting Disparate E-mail Systems

This chapter provides examples of the issues you enounter when you interface two or more e-mail systems, as well as a checklist for evaluating gateway products and vendors. Understanding how to interface systems and evaluate gateways is crucial for constructing a highly reliable and interoperable e-mail environment.

Almost any gateway product will suffice to interconnect multiple e-mail systems if you are not particular about message fidelity across the interface and do not care about the amount of effort required to manage the gateways and e-mail systems. But what if you or your organization wants to move messages between systems with very little loss of information and message formatting? What if your users expect a consistent, readily available system? In this case, the designing, choosing, installing, and managing of an integrated e-mail system becomes a moderately to highly complex undertaking.

The state of the industry is such that there are no widely used interoperability tests for gateway vendors or, for that matter, e-mail system vendors. E-mail interoperability is still an art, not a science. Basically, managers and administrators agree on the level of functionality required in the organization and then have selected vendors demonstrate whether their systems can achieve this.

NOTE: There is a consensus information-gathering process, called Reguest For Comment (RFC), *used on the Internet in which anyone can participate and provide input on particular issues being addressed. Two good documents that explain what is involved with using gateways to connect two e-mail systems are RFC 1327 and RFC 1494, and they are in the public realm on the Internet. They deal with how to map information between the Multipurpose Internet Mail Extension (MIME) message format body parts (discussed in Chapter 11) and X.400 format body parts (discussed in Chapter 12). (MIME is an enhancement for Simple Mail Transfer Protocol (SMTP), the e-mail format used on the Internet and UNIX-based computer systems.) These documents also define the relationship between envelope and heading fields in Internet and in X.400 messages. Together they give a very detailed view of what the issues are concerning messaging between different e-mail systems. Understanding these documents requires a fairly detailed knowledge of Internet and X.400 systems.*

Overview

Even though standards are well defined in many areas, there are no widely accepted interoperability tests to verify that a gateway meets those standards or that information is being converted in both directions in a standard manner. Also, the current standards are vague, and different companies implement them in different ways. Adding to the confusion is the fact that there are no functionality standards for all e-mail systems. Because of this current state of the industry, it is important to be cautious and verify functionality before implementing any gateway.

Some of the more apparent complexities involved in developing gateways are:

- Information mapping between e-mail systems can be nonsymmetric.
- Data structure types in one system might not exist in the other.
- Addresses are not always easily mapped between e-mail systems.
- Messages converted to another e-mail system and then reconverted back to the original system might not contain the same information as the original messages.

Nonsymmetric Information Mapping

Let's take a fairly straightforward example. Presume that the Subject field in e-mail System A is 80 characters in length and the Subject field in e-mail System B is 132 characters in length. Users of both systems use the complete length of the Subject field when sending the monthly financial data report. For example, a user from System B enters the following into the Subject field:

The ABC Tool Company's March 1995, Financial Report, Division Wrenches, Limited Distribution, TEST COPY

The above is more than 80 characters and will not fit in System A's Subject field. It is truncated when going in this direction. However, if System A sends a message to System B, the Subject line will not be truncated.

How do you handle the situation? Chop off the information, or put it somewhere else in the message? This is a simple example of nonsymmetric mapping, which gateway programmers frequently face.

Several examples of this problem exist in the X.400 and Internet gateways. One of these is the X.400 envelope field "DL Expansion Prohibited." Its purpose is to stop the expansion of hidden distribution lists. Comparable functionality does not exist in the Internet standards. Therefore, the request not to expand a distribution list is lost for the X.400 messages as they are converted to Internet messages.

Address Mapping Between Systems

Address mapping involves establishing a set of address values for one system that correspond with the values used by another system. Address mapping between disparate e-mail systems is a somewhat inaccurate, hit-or-miss affair, making it a major connectivity problem. Mapping uses algorithms that take an address apart and reassemble it in a different format.

To resolve this problem, tables are built that relate the actual address information for System A and System B. Address tables do not translate addresses—they actually contain the addresses. For example, when an Internet message arrives addressed to s.smith@abctool.com at a gateway that converts messages to Microsoft Mail, the gateway looks in a table for s.smith@abctool.com and gets the corresponding Microsoft Mail address sam.smith@finance.abctool.

Aside from constructing address tables, the inter-e-mail system address problem might be solved in the future by creating new, single, system-independent e-mail addresses. The latter is currently being discussed by various organizations as the need to simplify and reduce the number of addresses grows.

Information Discrepancies in Inverse Mapping

Here is an interesting situation. Suppose System A is an X.400 system, System B is an Internet system, and System C is an X.400 system. A user on System A sends a message to a distribution list on System B. The distribution list expands, with one of the distribution list addressees being a user on System C. The X.400 formatted message (System A) is translated to Internet format (System B) and then sent to another X.400 system where it is translated back to X.400 format (System C). Very often the resulting message is not in all ways the same as the initial X.400 message. This is an example of the original message and the resulting message not being the same when inversely mapped, or converted back and forth between transport facilities.

X.400 and Internet Message Mapping

The following presents issues involved with moving messages between different e-mail environments. The mappings in the exchange of messages between the X.400 (1988 version) and SMTP/MIME environments are used as an example. While many of the specifics are not necessarily applicable to other gateways between other products, the same general issues often exist, are relevant, and worth a closer study.

Before reading further, you might want to refer to Chapters 11 and 12 for more details on the envelope, message header, and message body of the X.400 (1988 version) and Internet environments. In the Internet environment the envelope and heading fields are combined, while in the X.400 environment the heading and envelope fields are separated into different parts of the message.

Heading and Envelope Mappings

Table 7-1 shows the X.400 envelope and heading fields and their relationship to the Internet RFC 822 envelope and heading fields.

Many field conversions are symmetric in nature. To, From, BCC, and Subject conversions are mostly symmetric in nature and exchange equally well in either direction, most of the time. For example, the Subject field in X.400 might contain characters that are a subset of the ASCII character set, while the Subject field in Internet would allow some additional printable ASCII characters such as "@," "%," and "_," to name a few. If the Internet Subject field contains "@," "%," and other characters, these must be encoded in a special way in the resulting X.400 message. For example, Internet Subject field *abcd@efg%* would convert to X.400 Subject field *abcd(a)efg(p)*, where *(a)* represents "@" and *(p)* represents "%."

X.400 Envelope/Heading	Internet RFC 822 Heading Fields
Date	Date
IPMIdentifier	Message Identifier
MTAIdentifier	X400-MTS-Identifier
Alternate Recipient Allowed	Not supported
Auto-Forwarded	Auto-Forwarded
Conversion Prohibition	Partially supported — IA5 only
Delivery Notification	Partially supported
DL Expansion Prohibited	Not supported
Forwarded IP Msg. Indication	Partially supported
Incomplete Copy Indication	Incomplete-Copy
Non Receipt Notification	Not supported
Originator	Either the From or Sender field
Copy Recipients	CC
Related IPM	References
Heading Extensions	Keywords
Primary Recipients	To
Blind Copy Recipients	Bcc
Authorizing User	From usually, could be Sender
Subject	Subject Partially
Replied to IPM	In Reply to
New Body part	Comments
Header extension	Encrypted
Header Extension	Resent
Obsoleted IPM	Obsoletes
Expiry Time	Expiry-date
Reply Time	Reply-by
Reply Recipient	Reply-to
Importance	Importance
Sensitivity	Sensitivity
Auto-forwarded	Autoforward
Languages	Languages
None	Discarded-X400-IPMS-Extensions

Table 7-1. *X.400 and Internet RFC 822 Envelope/Heading Fields*

So the Subject field, one of the easier fields to understand, is only partially symmetric in the X.400 and Internet gateways. Others are not symmetric at all, such as the Internet Comments field. If a message is being converted from Internet to X.400

and has a Comments field, then the comment data is put in an X.400 new body part and identified as a comment. The inverse mapping does not exist.

Some X.400 heading fields are tossed away as X.400 messages are converted to the Internet. A special field, called Discarded-X400-IPMS-Extensions, can be used to show deleted heading fields in the resulting Internet messages.

Adding to the complexity is that X.400 defines different types of e-mail messages, while the Internet defines only one. Below is a paraphrased quotation from the author of RFC 1327, S. Hardcastle-Kille, of the University College, London:

> When transmitting an RFC 822 message to an X.400 system, the RFC 822 message and the associated transport information is always mapped into an Interpersonal Messaging (IPM) message. In transmitting an X.400 message to an RFC 822 system, an RFC 822 message and the associated transport information may be derived from:
>
> 1. An MTA report
>
> 2. Interpersonal notification (IPN) such as delivery and nondelivery
>
> 3. An IPM e-mail message itself

Transport information contains the routing instructions and heading information necessary to transmit the message through the Internet. The Internet system builds this information by using fields taken from the X.400 MTA report, a delivery or nondelivery notification, or from the X.400 message itself.

Body Part Mapping

We have looked at the various ways the heading and envelope parts of the message are mapped back and forth between Internet and X.400 messages. Now take a look at how the third major part of an e-mail message, the body, is mapped between X.400 and Internet MIME. In Table 7-2, the X.400-to-Internet MIME and the Internet MIME-to-X.400 general content and body part mappings are listed. Much of the table is taken from RFC 1494. This table shows the difficulty of sending messages from an Internet system to an X.400 system and vice versa. Most of the content types are self-explanatory. Note that sometimes the mapping works both ways, sometimes the mapping varies slightly, and sometimes there is no applicable body part mapping at all. For example, Group 3 facsimile content types work the same each way, whereas for Group 4, no type is defined on the MIME side. In the case of the latter, a user could send a Group 4 fax via an X.400 system, but it would not be fully translated into an Internet MIME message.

NOTE: *In Table 7-2, EBP stands for extended body part. Unlike basic body parts, extended body parts are externally defined, meaning that they are not specified in the X.400 standard, such as a Microsoft Word for Windows 6.0 word processing document.*

MIME-to-X.400 Mapping

MIME Content Type	X.400 Body Part
text/plain	IA5-text
charset=us-ascii	ia5-text
charset=iso-8859-x	EBO-GeneralText
text/richtext	No mapping defined
application/oda	EBP-ODA
application/octet-stream	bilaterally-defined
application/postscript	EBP-mime-postscript-body
image/g3fax	g3-facsimile
image/jpeg	EBP-mime-jpeg-body
image-gif	EBP-mime-gif-body
audio/basic	No mapping defined
Video/mpeg	No mapping defined

X.400-to-MIME Mapping

X.400 Body Part	MIME Content Type
ia5-text	Text/plain;charset=us-ascii
voice	No mapping defined
g3-facsimile	image/g3fax
g4-class1	No mapping defined
teletex	No mapping defined
videotex	No mapping defined
encrypted	No mapping defined
bilaterally-defined	Application/octet-stream
nationally-defined	No mapping defined

Table 7-2. *Conversion Table for X.400 Body Parts and Internet MIME Content Types*

Gateway Functionality Verification

Planning an e-mail integration project requires an understanding of the general issues involved in interfacing different e-mail systems. This section focuses on a list of functionality tests to assist in the gateway evaluation process. These tests are of paramount importance in ensuring that messages are not lost or distorted as they are converted between systems.

Functionality is the features, services, and limitations of a gateway in enabling successful exchange of messages across disparate systems. Lost messages and information distortion within messages create problems when systems are interfaced across gateways. You might want to refer to Chapters 4 and 15 for information on gateway vendors before reading further.

Prior to purchasing a gateway to interconnect disparate e-mail systems, it is important to perform a thorough evaluation of the gateway's functionality. This can be done either by running a pilot on your premises using test data or by working with the vendor at the vendor's location. Either way, these tests are vital in order to ensure that the gateway meets all the message conversion standards and your performance specifications. This section contains descriptions of the particular services to be tested for acceptable functionality, as well as suggested test scenarios. *Services* are defined in the X.400 standard as features, functions, or capabilities that enable users to communicate with each other using an X.400-compliant messaging system.

Service: Address Mapping

One of the more difficult parts of e-mail integration is mapping addresses between e-mail systems. Bob Denny, President of Alisa Systems, Inc., says the real test of any gateway is to see if "reply all" works by replying to all users on the original address lists. A gateway should map all addresses in the To, CC, Sender, BCC, and Reply-to fields so that each user receiving the message can respond to the whole address list without additional work. Most e-mail systems have either a function key or an icon for the reply or reply all function that enables a user to automatically reply to the message with the touch of a key. The e-mail system then creates a new message, plugging in the return addresses that appeared in the original message. In the new message, the addresses should be accurate and valid so that a reply can easily be sent back to the originator.

TEST: The only way to test the address mapping functionality of different gateways is to issue a series of single messages to individual users on each possible destination system, and record those users' ability to respond to all users on the original address list. This test should cover all address fields and include distribution list expansion as part of the test.

Service: Character Set Mapping

E-mail systems can use different character sets. Sometimes different character sets are used even within the body and heading fields within the same e-mail system. Address strings are very sensitive to the character sets used.

TEST: Enter a full character set in different portions of the message body and heading fields, and verify that the conversion takes place appropriately.

Service: Limitations on the Number of Addresses

Most systems have a limit on the number of addresses they can list. This feature prevents a user from clogging the messaging system by addressing a message to thousands of recipients. For some systems, the limit is very high; for others, especially older systems, it is lower.

TEST: Enter an appropriate number of addresses for your system's application and verify that all addresses are handled correctly by the destination systems.

Service: Subject Field Length

Subject fields have different lengths and are measured in numbers of characters.

TEST: Verify that the appropriate sized subject field is supported by entering very long strings and checking the length of the destination subject fields. For example, you could enter a string of the following form:
    1234567891123456789212345678931234567894...

and then check to see how much of the string carried over to the destination subject field.

Service: Types of Time Stamps

Several time stamps exist in e-mail messages. Many systems use the UTC time which is in the form *yymmdd hh:mm -/+xxxx*, where *xxxx* is the time set from coordinated universal time, as established by the Utilities Telecommunications Council (UTC) in Greenwich, England. The western hemisphere from UTC time is a negative offset; the eastern hemisphere from UTC time is a positive offset.

TEST: Record time stamps of source messages and verify that each individual time stamp is translated properly in each destination system.

Service: Subject Field Contents

Not all systems support the same character set in the Subject field. This service determines which symbols and characters will be successfully carried in the Subject field to different e-mail systems.

TEST: Enter ASCII characters and record those that convert and those that do not. For example, be sure to test symbols such as @, #, and $.

Service: Message Unique ID/Trace string

Each message often has two unique identification strings. The first is part of the submission, transport, or delivery envelope, and the second is part of the e-mail message heading. These are both key fields, because they are necessary for implementing some rule-based message sorting systems and tracing messages for problems.

TEST: Create a message that records the value and structure of the message identification strings. Verify that the strings are traceable from the original message by tracking the message from the originating system all the way to the recipient system.

Service: Priority

Most systems have some way to assign a delivery priority to a message. Low, Normal, and High are the usual types of priority systems supported.

TEST: Assign different priorities to messages and verify that the destination messages have comparable priorities after gateway conversion.

Service: Content Information—Body Part/Content Type

In some message headings, the type of data (ASCII, Rich Text Format, and so on) contained in the message body parts is recorded. These recordings are often called *content identifiers*. They can be lost during conversion.

TEST: Content identifiers are difficult to test without using messaging APIs to analyze the message heading. If possible, verify that different body content types are carried end to end.

Service: Alternate Recipient Allowed

Some systems provide this service, which automatically forwards messages, at the destination user's request, to other users. Users set the autoforward option when they go on vacation or when they want to forward messages to a designee. When this field is set, it means messages can be autoforwarded.

TEST: Set the field in the source system, send a message to an account that is autoforwarded, and then verify that the message was forwarded.

Service: Autoforwarded Indication

This field contains information that indicates that the message has been forwarded to the destination account.

TEST: Send a message to a local account that is autoforwarded to an account across the gateway. Verify that there is some sort of autoforward indication in the forwarded message.

Service: Blind Copy Recipient Handling

This service ensures that blind copy recipients are not disclosed to primary recipient addressees.

TEST: Send a message including To, CC, and BCC addresses, and verify that the To and CC recipients cannot see the names of the BCC addressees or any indication that a blind carbon copy was sent to anyone.

Service: Body Part Encryption Indication

This service provides an indication that the message contains encrypted body parts. This service is not often seen across multiple systems at this time, so no tests have been developed for it.

Service: Content Type

This service shows the types of information contained in the body of the message. Information such as Group 3 fax and Microsoft Word version 5.0 are examples of information contained in this field.

TEST: Verify that the particular information type converts appropriately across gateways by testing the conversion of the main information types your organization is using.

Service: Conversion Prohibition

This service prohibits the conversion of message body parts.

TEST: Set this indicator, send messages containing the different types of body part information that your gateways are able to convert, and verify that conversion does not take place on those information types.

Service: Conversion Prohibition in Case of Loss of Information

This service prohibits conversion if loss of information will take place. For example, in the case of a word processing document, if features such as bolding or underlining will be lost, the body part will not be converted.

TEST: This service is so specific to the document conversion facility that it is very hard to test. Try creating a test message with a significant amount of text features, such as bolding and highlighting, and send it through to the other system to see if the features convert.

Service: Converted Indication

This service provides an indication that conversion has taken place on a document-by-document basis.

TEST: Send a message with an attached document through a gateway that will convert the attachment, then verify that it is converted and that the conversion is noted in the delivered message.

Service: Cross-referencing Indication

This service utilizes a heading field that allows users to "tie" together the body parts with a description in this field.

TEST: Create a test message composed of three different body part types. For example, use a Group 3 fax, an ASCII file, and a text message. Select the cross-referencing service, send the message, and verify that the received message indicates that the three body parts are tied together.

Service: Delivery Notification

A delivery notification indicates that the source message is requesting the remote delivery MTA to notify it that message delivery has taken place. This does not mean the message has been read; it only means it has been delivered. Some systems support delivery, nondelivery, and read receipt notifications; others support only a subset of these. It is important that each notification type be mapped to the appropriate message as it moves across the gateway.

TEST: Send a message requesting delivery notification across the gateway to each different system. Verify that notification is returned and that the notification identifies the original message so that the message and the delivery notification are paired.

Service: Read Receipt

This service is not supported across all systems and should not be assumed in an implementation. Unlike delivery and nondelivery notification, which are created by actions of the transport system, the read receipt is created by the remote MUA. If read receipts are supported by the destination MUA, the user normally has a choice of whether or not to allow them. Most users see read receipts as an intrusion and normally will not issue them.

TEST: Send a message to an end user whose e-mail system is capable of sending back a read receipt, and then verify that a read receipt is received by the originator.

Service: Nondelivery Notification

This service provides a report that the message could not be delivered to the address.

TEST: Send a message to a nonexistent user and verify that a nondelivery notification is returned identifying the corresponding source message in an ID field.

Service: No Nondelivery Notification

This service requests that no nondelivery messages be sent, even if a message cannot be delivered. This service is used for messages with large distribution lists.

TEST: Send a message to a nonexistent user on another gateway system, indicating that no nondelivery message should be returned, and verify that you do not receive a nondelivery notification for that message.

Service: Return of Content with Nondelivery Notification

In some systems, a user might request that the entire body of the message be returned as part of a nondelivery notification, to help match the nondelivery with the original message.

TEST: With this service indicated, send a message across a gateway to a nonexistent address. Verify that the message content is part of the nondelivery message.

Service: Delivery via Other Distribution Means: Fax, Telex, or Postal

This service provides for message delivery via nonelectronic means. It is often used to deliver information to people outside of the organization who are accessible only by fax, telex, or postal means. Messages delivered through nonelectronic means are not always in the proper format; for example, fonts, positioning, and letterheads often are not displayed as anticipated.

TEST: Send messages to each type of delivery mechanism and verify that the output matches what is expected in areas such as fonts, letterhead, positioning, and so on.

Service: DL Expansion History Indication

Distribution lists are expanded by the MTA as the message moves across the network to the destination. This service shows the history of distribution list (DL) expansion in the destination messages. For example, a distribution list might contain an address that is a distribution list and also contains other distribution lists.

TEST: Compose a message and address it to a distribution list that contains individual addresses of recipients on the source system and to an address that is a distribution list of recipients on the destination system. Verify that the expansion of the distribution list is recorded properly.

Service: DL Expansion Prohibited

Distribution list expansion can cost money on some services. The user should be able to specify that distribution list expansions should not take place.

TEST: Set this field, include a known distribution list that will be expanded on the other system, and verify that the distribution list is not expanded.

Service: Express Mail Service (EMS)

Most systems allow one priority to be an express or high priority, which causes a real time transfer of a message to the destination. The gateway should recognize this and transfer the message to the other system ahead of lower-priority messages.

TEST: Send a high priority message across the gateway, and verify that the gateway and destination system handle the message before lower-priority messages.

Service: Multipart Body

Most systems support the inclusion of multiple body parts.

TEST: Send a message with several different body part types and/or an attachment, to the destination system to verify that it is handled as anticipated.

Problem Resolution

If any of the preceding tests fail during the pilot or the on-premise evaluation at the vendor location, you have several options. First, determine whether the particular services that failed are important to the organization. For example, if your company does not permit read receipts, and this test failed, obviously the result can be overlooked. Second, determine the extent of the failure. Are all or just one of the participating e-mail systems experiencing the problem? Isolate which systems are not capable of handling the particular service. Is the problem confined to the gateway itself? Third, based on the nature of the problem, you can negotiate with the gateway or e-mail system vendor to fix the failure by issuing a software change. Often, the failures that one customer experiences are also happening to others. The vendors then generally offer a new release to resolve the problems. If time is a constraint, you must consider how much time the vendor will take to release a new version of the product.

Fourth, if the vendors cannot or will not fix the problem, and this service is critical to you, you must stop the evaluation of this gateway product and look for an alternative product to perform the e-mail system integration. Alternatives might be using a different point-to-point gateway, using a system based on the X.400 standard, or using a value added network (VAN) to handle the system interconnection.

Summary

E-mail system interconnections are still ill defined and require that the system architect make sure system functionality is investigated before implementation. Implementing the MUA, message store, and MTA services specified in the X.400 1988 and 1992 standards often forms a good set of e-mail services.

PART THREE

Electronic Messaging Concepts

Chapter Eight

X.500 Directory Systems

It is simple to locate anyone's telephone number in your local area by dialing 411, or in the United States by dialing the area code and 555-1212, or in the world by calling an international directory assistance operator. This capability of finding almost anyone's telephone number fostered enormous growth in the telecommunications industry, and switched universal telephone service from a dream to a reality.

No common addressing convention such as the ten-digit telephone number exists in the electronic messaging arena, nor is there an easily accessible worldwide electronic directory facility. This capability, this global electronic "411," is the concept behind the X.500 directory standard.

In 1988, the Consultative Committee for International Telegraphy and Telephony (CCITT) published a set of recommendations for developing standardized software for a global, distributed directory that would facilitate the exchange of electronic information over data communications networks. The standard (X.500–X.521) describes the structure of the directory information model, the protocols necessary for interoperability between cooperating open systems, and procedures for distributed operations.

The X.500 directory standard is updated every four years. The most recent version was released in 1993, and already vendors have incorporated it in their directory product offerings. Microsoft's Electronic Messaging System (EMS) is based on X.400 and X.500 and is scheduled for release in late 1994. Lotus Development's Lotus Communications Architecture (LCA) products are also based on X.400 and X.500 and will be out in mid-1995. Mainframe vendors such as DEC, Hewlett Packard, and NCR also offer X.400 and X.500 products.

Overview

The X.500 directory architecture is intended to be a worldwide, fully distributed directory that supports retrieval of e-mail addresses, telephone numbers, and individual postal addresses. It is no different in concept from the series of cooperating distributed databases that are common from vendors such as Oracle, IBM, and SyBase. Often the X.500 directory is implemented using these types of products as the database. The X.500 standard differs from such products in that it defines in a non-vendor-specific manner how these distributed databases communicate: how the information, components, and services are organized and how the components intercommunicate.

Wide scale implementation of X.500 is several years away. Many companies are implementing X.500 on a single database instead of multiple distributed databases to make implementation and management less complex. As we move toward a global electronic marketplace, an X.500 directory system will benefit all organizations, regardless of size, by providing consistent and easily accessible information.

Originally developed at the University College, London, the X.500 directory system software, called QUIPU, used a UNIX operating system and a SUN workstation as its hardware platform. This software is now in the public domain and is available on the Internet, free of charge. Software can also be purchased as part of an e-mail or gateway system from numerous vendors such as DEC, Soft*Switch, Control Data Systems, and Isocor. Value added networks (VANs) such as MCI and PacTel are also beginning to offer X.500 directory services for their subscribers.

The implementation of X.500 encompasses several areas: the technical implementation of the databases, the definition of the structure of the information contained in the database, and the implementation of supporting policies and procedures necessary to ensure that the data in the X.500 database is valid and regularly updated.

Often the most difficult aspect of adopting an X.500 system is the politics of information within the implementing organization. X.500 requires a company to combine information contained in the company phone book, personnel database, and e-mail address database in one place—the X.500 directory. At that point, information ownership and responsibility start to blur causing political blocks and controversy. Combining these diverse areas means that the structure of the database must be designed to support the different types of information. Many types of information structures can be defined; in fact, a whole book could be written on this subject alone. The Corporation for Open Systems (COS) is currently involved in implementing a large X.500 project and is documenting the process. Most electronic messaging vendors and consultants have processes in place to help a company get around the technical and political minefields involved in an X.500 implementation.

Alexis Bor, Boeing's X.500 project manager, suggested the following steps when starting an X.500 project:

1. Identify reliable information sources for loading the X.500 directory, and validate information update processes. Information sources such as personnel databases, vendor databases, phone books, and e-mail directories are all relevant.

2. Understand the nature of your data, how it will be used, what future changes are anticipated, the stability of your organization (both organizational and geographical), and analyze any constraints that may be placed on the data such as security, availability, and accessibility.

3. Design your directory tree structure.

4. Develop an appropriate database schema to ensure that data is entered and maintained with correct values.

5. Develop an access control strategy for both internal and external access to the database and the individual pieces of the data. Since the X.500 directory entry can be robust, the organization may not want specific information contained in an individual's entry to be viewed either by internal employees or by users from outside the company. For example, information such as a user's department and title would be very useful to executive recruiters. The directory can be structured to deny such outside requesters access to this information.

6. Establish a production commitment for providing 24-hour availability across all distributed directories.

7. Make directory user interfaces available for the desktop by site licensing vendor products, or develop your own. Directory user interfaces should support at least Windows, Macintosh, UNIX, and DOS desktops, so that users can query the X.500 database for information.

8. Work with e-mail vendors to encourage integration of proprietary directories into X.500. Each e-mail system today has its own proprietary directory system that must work with the X.500 directory, at least for synchronization, until it is converted to the X.500 standard. Vendors and end users in the e-mail industry are now working to develop a standard for this purpose.

9. Make sure that directory synchronization tools are available for synchronizing existing proprietary e-mail directories with X.500 DSAs.

The X.500 directory can also support commercial yellow pages, listing businesses and their product offerings. It may hold information on network addresses, network devices, and other network services. It is a generic directory with a hierarchical tree architecture that supports a wide variety of information types. The X.500 directory is specified by the International Telecommunications Union in its ITU-T X.500 series and the International Organization for Standardization (ISO) in its ISO/IEC 9594 standard recommendations to support multiple telecommunications systems.

Although X.400 and X.500 sound similar and were developed by the same international standards organization, they are not similar in purpose or design. X.400 is a transport protocol for data communications networks. X.500 is a model for a distributed directory facility. X.400 does not require X.500 to function, and X.500 can operate independently of an X.400 network. In fact, Internet currently has one of the largest X.500 implementations in the world supporting non-X.400 users.

This chapter focuses on X.500 directory use in the electronic messaging environment. The X.500 directory specification was initially prompted by the need for a directory in the X.400 world. It quickly became apparent that many telecommunications system components also could be helped by a generic directory, so the CCITT defined a directory with the ability to support many technologies. Because of the close association of the X.500 and X.400 teams, X.400 uses X.500 functionality in four areas:

■ *User friendly X.400 addresses*   The 1988 X.400 and X.500 standards enable users to enter a directory name—a more friendly form of the X.400 address—to address a user. When the X.400 system needs to find the address of a specified user, it requests the detailed address from the X.500 directory. At the moment, most X.400 products do not directly use the X.500 functionality for routing; they use X.400 originator/recipient (O/R) addresses for that purpose.

- *Directory user agent (DUA) supported functionality description* Directory user agents (DUAs) have a range of capabilities as to the amount of information they may hold. The directory may be used by X.400 to specify the maximum size of a document receivable by a message user agent (MUA) and the type of functionality supported. Information contained in the directory will be the basis for automatic document conversion services, electronic signatures, and least-cost routing preferences. Full implementation of these features in most X.500 installations is several years away.

- *Authenticity and security* Both X.400 and X.500 offer two kinds of authenticity and security: simple and strong. The simple method uses a common password technique for validation. The strong method uses tokens that are normally based on asymmetric key encryption techniques. Key systems have two keys for each user: one is public and the other is private. The public key must be widely accessible to messaging users for authenticity and signature schemes to be effective. The public part of the key is kept in the directory and used by the electronic messaging user to verify who sent the message. The private part of the key is known by only one user.

- *Distribution lists* A distribution list (DL) contains a list of e-mail addresses or other distribution lists. These are already contained in the directories of many proprietary e-mail systems. When a message transfer agent (MTA) encounters a DL address, it queries the directory for an expansion into its components. New addresses are then expanded and added to the message.

X.500 Directory Components and Protocols

The X.500 architecture has two main functional components, directory user agents (DUAs) and directory systems agents (DSAs), and three protocols—Directory Access Protocol (DAP), Directory System Protocol (DSP), and Directory Information Shadowing Protocol (DISP) discussed in the following list. The user, program, or individual accesses the directory through the DUA using DAP. DSAs use DSP to intercommunicate.

DIRECTORY ACCESS PROTOCOL (DAP) DAP is an application protocol used between the user interface (DUA) and the database (DSA). It enables the user and other directories to remotely query the X.500 directory.

DIRECTORY SYSTEM PROTOCOL (DSP) DSP is used between DSAs to specifically support the referral, chaining, and multicasting between interacting DSAs. These types of interactions are initiated by a user (DUA) information request and are discussed in more detail later.

DIRECTORY INFORMATION SHADOWING PROTOCOL (DISP)   DISP is used to share copies of information between specified DSAs. Unlike the DSP protocol, which is used to request and transfer user requested information between DSAs at the time of the request, DISP is used to copy and maintain entire portions of the database on other DSAs to enhance search requests.

ADMINISTRATIVE DIRECTORY MANAGEMENT DOMAIN (ADDMD)   ADDMD contains from one to many DSAs, a directory system all under one management authority—usually commercial in nature. In Figure 8-1, Japan is an ADDMD.

PRIVATE DIRECTORY MANAGEMENT DOMAIN (PRDMD)   PRDMD contains from one to many DSAs. ABC Tool is an example of a PRDMD in Figure 8-1. ABC Tool Company's PRDMD contains only one DSA and is not managed for commercial service; it may choose to share or not share directory information with other PRDMDs or ADDMDs.

Each DSA contains part of the database, or *directory information base (DIB)*, as it is called in the X.500 specification, which is the information in the worldwide directory. In Figure 8-1, the ABC Tool Company DSA, a database residing on a single computer owned by ABC Tool Company, contains the demographic data for the employees and departments within the company. The type of information contained is very much like that contained in IBM's CallUp or Enterprise Address Book, or Microsoft Mail's Global Directory. The demographic data held by ABC Tool Company's DSA are office and home postal addresses, e-mail addresses, phone numbers, titles, full names, and nicknames.

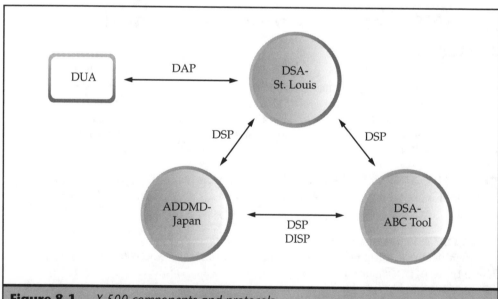

Figure 8-1.   *X.500 components and protocols*

Directory Information Tree (DIT)

The X.500 directory is built on distributed database technology. It has the hierarchical structure of a tree with a root. Directly under the root are countries, with further divisions being defined by the country's ADDMD and PRDMDs. In the example of a directory information tree (DIT) shown in Figure 8-2, both St. Louis and ABC Tool belong to the portion of the tree under U.S.A. St. Louis has many entries under it, with only three showing in the example. ABC Tool is a nationwide company with its main offices in St. Louis. ABC Tool is registered twice in the tree. The main entry is directly under the U.S.A leaf, with an "alias" entry under St. Louis pointing to the main entry under U.S.A. Aliases help shorten the search time for a particular entry. A requester, with knowledge that ABC Tool is located in St. Louis, could reduce the number of displayed entries by entering St. Louis in the search criteria instead of U.S.A.

The alias entry usually has no additional information of its own. It only points to the main entry where the information is kept. All information about ABC Tool Company contained in its directory database in St. Louis is tied into the national U.S.A. ADDMD. For example, the U.S.A. ADDMD would direct a request from a user in Georgia for information on an ABC Tool employee to the ABC Tool DSA in St. Louis to retrieve the information. Figure 8-3 shows the ABC Tool single DSA tree, which, for the purpose of simplicity, has only two branches—Personnel and Departments.

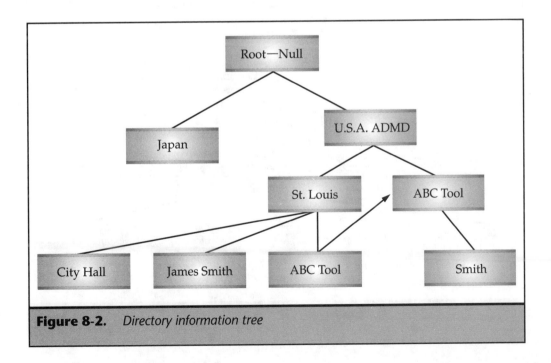

Figure 8-2. *Directory information tree*

Distinguished Name

A *distinguished name* is the unique name of a database entry. The purpose of the distinguished name is to create an unambiguous name that has the capability of being unique in the world. It is made up of the sequence of the superior nodes in the directory tree. In Figure 8-3, reading down the tree from top to "Alice" at the bottom, Alice's distinguished name would be

ABC Tool, Personnel, Smith, Alice

She is found under the U.S.A ADDMD, under ABC Tool, under Personnel at ABC Tool, under Smith with the name Alice.

If there were two Alice Smiths at ABC Tool, some of the attributes of the "object" Alice would be used as "distinguished attributes" to create the unique, distinguished name. The "Director" attribute would be the tie-breaker between the two entries.

Director, Alice, Smith, Personnel, ABC Tool, U.S.A.

The entry "Alice" has several pieces of information, or *attributes*. Five attributes are shown in Figure 8-3.

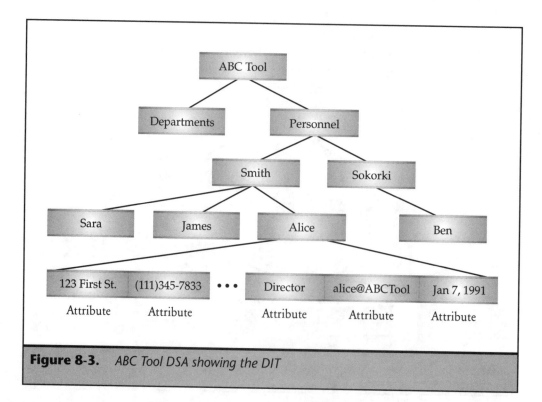

Figure 8-3. *ABC Tool DSA showing the DIT*

Access from the DUA

The DUA is used by individuals and programs to communicate with the DSAs. This is analogous to an SQL request to an SQL database. The protocols are different, but the logic is the same. The dialogue between the user (DUA) and the database (DSA) may be of two general types: queries and administrative. This chapter focuses on query dialogues.

Directory Query

There are five types of query requests that can be made by the user to the directory database: read, compare, list, search, and abandon.

READ   Read returns the attributes of a particular single entry. For example, "Read Alice Smith." In this query, a user requests that the directory locate the specific entry for Alice Smith and display it on the user's computer screen. The directory locates the entry and returns all attributes such as title, phone, address, and e-mail address to the requester.

COMPARE   Compare returns a True or False from a comparison of a single attribute value of a single entry. For example, "Is Alice Smith's address 121 First St.?" If the statement is true, the user would see Alice Smith's directory entry. If not, the directory would return a message stating that no entries matching that criteria were found.

LIST   List returns a list of the subordinates of an entity. For example, "A list on Smith" would return all the entries directly under Smith: Alice, Ted, Ben, ... James.

SEARCH   Search returns specific attributes of all entries found in a portion of the tree. The user, as part of the search criteria, specifies what attributes, values, and types to return and the portion of the tree to search. For example, Search Country = U.S.A., Company = ABC Tool, for all personnel, and return First Name, Last Name, and Title. The directory returns a list of entries with those specified attributes to the requester. The list contains the full names and titles of all the employees of ABC Tool within the U.S.A.

ABANDON   Abandon stops a search in progress. Sometimes very complicated or lengthly searches can become costly and this function enables a user to terminate the requested search.

DSA Cooperative Queries

The X.500 standard identifies three types of inter-DSA cooperation necessary to resolve user queries. They are chaining, multicasting, and referral.

CHAINING   The DSA directly servicing the user (DUA) passes the query on to another DSA to fulfill the request on behalf of the first DSA, as shown in Figure 8-4.

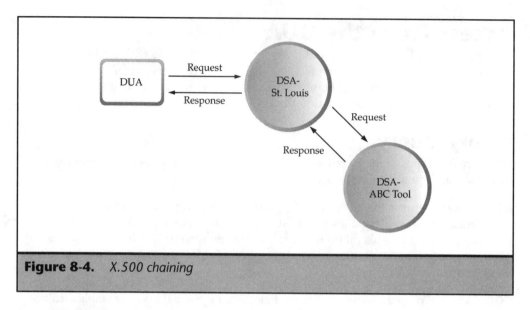

Figure 8-4. *X.500 chaining*

Several DSAs may pass the request on until a DSA with the necessary information is able to resolve the request.

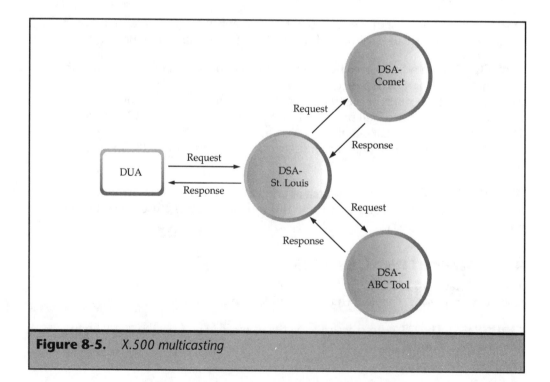

Figure 8-5. *X.500 multicasting*

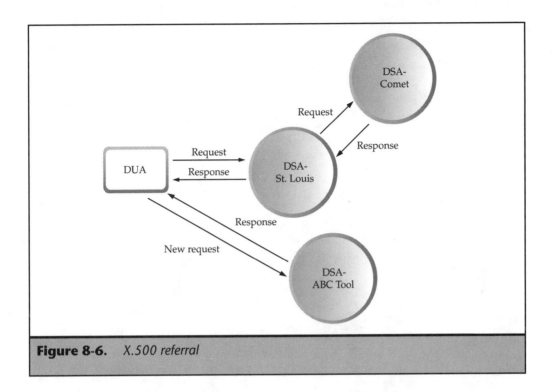

Figure 8-6. *X.500 referral*

MULTICASTING   A user (DUA) request is passed on simultaneously to multiple DSAs, looking for the one that can resolve the request. Those that cannot return an Unable to Process Service Error. In chaining, the request is passed down a single sequence of DSAs, while in multicasting, the request may be passed to several sequences of DSAs at one time. See Figure 8-5.

REFERRAL   A DSA or a DUA query request may cause a referral to be returned to the requester. Unlike the first two, where information is returned to the user, in referral mode the user is told to which DSA to go to get the information. See Figure 8-6. The user (DUA) then reissues the query to the designated DSA.

Summary

A few companies have implemented X.500, more are testing intermediate pilots, and many are planning for a future X.500 implementation. This robust directory model supports postal addresses, phone numbers, faxes, multiple e-mail addresses, cellular phones, car phones, telex addresses, public encryption keys, and much, much more. It can support millions and millions of users and many other commercial uses will be discovered and implemented as the technology matures. At that point, the vision of X.500 as the worldwide electronic directory will have become a reality.

Chapter Nine

IBM SNADS and DIA

Systems Network Architecture Distribution Services (SNADS) and Document Interchange Architecture (DIA) are IBM's multiplatform electronic messaging products. They resemble X.400 more closely than they do SMTP and Novell's NetWare Global MHS, two other predominant systems. SNADS and DIA are comparable to X.400 in the way the envelope and documents are encoded and in the way messages are moved between message stores. E-mail gateway vendors dread writing the interface to SNADS and DIA because the environment is complex and the names and acronyms are difficult to master, not to mention the concepts. This chapter compares SNADS and DIA to three multiplatform architectures: X.400, SMTP, and Novell Global MHS.

IBM's Messaging Architecture

The e-mail messaging architecture for the IBM environment is a mix of products. The three main offerings are OfficeVision/VM (OV/VM), OfficeVision/MVS (OV/MVS), and OfficeVision/400 (OV/400). The first, OV/VM (discussed in Chapter 6, "Description of Private E-mail Systems"), is a descendent of PROFS and is based on a messaging protocol called ZIP5. The last two are built on the SNADS protocol, which is the focus of this chapter.

SNADS is the mechanism that moves mail, transfers files, and distributes documents between e-mail systems. DIA is the way the messages and attached documents are structured for storage in the *file cabinet* (IBM's name for message store), for searching and retrieval, and for transport to different systems.

Addressing

Addressing in IBM SNADS products is based on two eight-character strings, with the first (left) string being the *distribution group name (DGN)* and the second (right) string being the *distribution element name (DEN)*. The format, called DGN/DEN (often pronounced "diggen den"), is similar to the PROFS Nodeid/Userid address and is used across all OfficeVision systems. The address may be that of a person or distribution list (DL). An example of a DGN/DEN address is

SYSTEM1:MARKGOOD

SYSTEM1 is the DGN eight-character string and address of the SNADS mail system computer, and MARKGOOD is the DEN eight-character string and name of the user Mark Good.

Transport

SNADS transports messages using a store-and-forward system and user agents called distribution service units (DSUs). The DSUs are similar to the X.400 MTAs, described in Chapter 12, "X.400 Interpersonal Messaging," in which the message, along with a list of requested services, is passed to the e-mail system that creates the envelope. The messages are transported over an SNA network that uses *network addressable units (NAUs)*—origin or destination points—to provide reliable transfer of data among end users. See Figure 9-1 for an example of message flow between SNADS and DIA. SNADS, like all message transports, is only used to move messages between different computers. Messages addressed to users on the same machine are never given to SNADS—they are usually delivered through the *Distributed Office Support System (DISOSS)*, an IBM product that enables distribution of information and translation of documents.

SNADS Message Structure

In SNADS the composite parts of the e-mail message are the message envelope, the message heading, and the message body form. In SNA terminology, these parts together are referred to as the *message unit.* The message unit contains a basic information unit (BIU), a path information unit (PIU), and a request/response unit (RU). This structure is similar to that found in the X.400 standard.

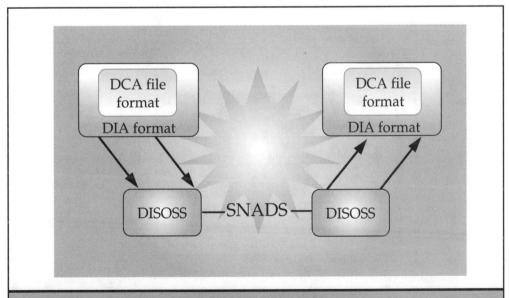

Figure 9-1. *SNADS/DIA Intersystem Document Exchange*

Message Envelope

In SNADS, the distribution interchange unit, alternatively called the distribution transport message unit (DTMU), and the Document Interchange Architecture (DIA) together comprise the entire message. The *distribution interchange unit* is the envelope routed through the SNADS network, and it contains a DIA document interchange unit file. This is similar to the P1 envelope containing the P2 contents in X.400. See Figure 9-2, which shows the structure of a SNADS distribution interchange unit.

Message Heading

The message heading and envelope fields are mixed in with the document interchange units of SNADS and the distribution interchange units of the DIA structure. They are not in distinct areas, as is the case in X.400 or Novell's Standard Message Format (SMF). This creates interoperability problems between SNADS and non-SNADS e-mail systems. Also, a SNADS gateway may need to be inserted into the e-mail integration architecture to translate from an SNA environment to another protocol such as X.400, TCP/IP, or DECnet.

Several e-mail system and gateway vendors have developed products suited to this purpose. For example, Emc^2/TAO offers a SNADS gateway with full support for *notes*, IBM's term for messages, and DIA documents; and cc:Mail uses the PROFS Link product for connectivity to the OfficeVision (PROFS) e-mail system via a SNADS

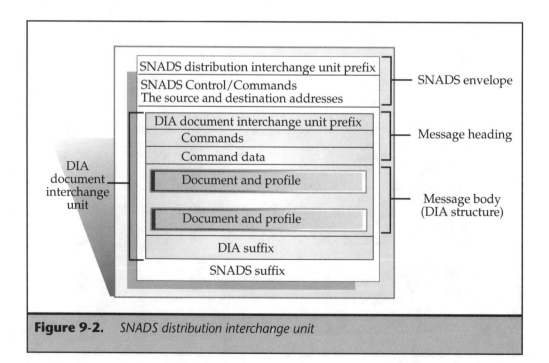

Figure 9-2. *SNADS distribution interchange unit*

gateway. DEC's Message Router SNADS provides connectivity between the SNADS environment and that of the DEC MAILbus system. Also, Soft*Switch markets a SNADS gateway for their EMX product, and Hewlett Packard offers the HP DeskManager DISOSS and SNADS gateways.

Message Body—DIA Structure

The message structure is coded in a manner analogous to, but not the same as, the Abstract Syntax Notation (ASN.1) encoding defined by the International Telecommunications Union (ITU). ASN.1 is used to describe the generic structure of the data, or *syntax,* used as the basis for the X.400 and X.500 standards. ASN.1 is discussed in Chapter 12, "X.400 Interpersonal Messaging."

The SNADS document, including the envelope, heading, and body, is transmitted between users inside a *document interchange unit (DIU)*. The DIU has the form shown in Figure 9-3. In the figure, the document(s) section includes the message body, and the command data section is the message heading.

The documents being interchanged have a specific structure. Each different part of the document, or individual information field (for example, the Subject field), starts with a five-byte segment descriptor called the *DIU introducer,* alternatively called the length identification format (LLIDF).

NOTE: *The introducer may optionally have three additional bytes that are appended. These are used when a document spans multiple transmissions (DIUs).*

The DIU introducer describes the segment length, content type, and format in the same way the ASN.1 encoding in the X.400 and X.500 standards does. The DIU introducer and the rest of the segment has the general structure shown here:

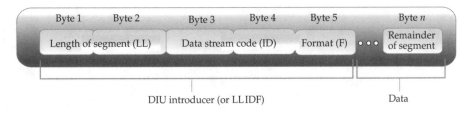

The first five bytes of the DIA segment are the DIU introducer; the remaining part of the segment is the data.

A DIA segment can be nested. For example, a DIA segment composed of a DIU introducer and data can contain a DIA segment composed of yet another DIU introducer and data. The structure is recursive in nature, as are SMTP/MIME body parts, X.400 ASN.1 encoding, and X.400 body parts.

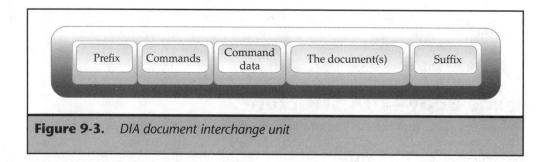

Figure 9-3. *DIA document interchange unit*

Address Directory

SNADS has no built-in directory but can use directories such as the Enterprise Address Book or flat files listing users and addresses. Most organizations develop their own in-house directories for their SNA networks. This presents interoperability problems in synchronizing with an X.500 directory system, as discussed in Chapter 8, "X.500 Directory Systems."

Summary

SNADS is the de facto transport protocol within the IBM environment. It provides reliable transfer and distribution of messages, documents, and files for mainframe IBM users. Most e-mail and gateway vendors have developed specific products to provide interoperability with SNADS, but development of such products is complex, requiring knowledge of SNADS' acronyms, syntax, and message format to achieve optimum interoperability with other messaging systems.

Chapter Ten

Novell NetWare Global Message Handling Service

Novell's NetWare Global Message Handling Service (MHS) is an open electronic message exchange protocol that usually runs on top of Novell NetWare servers but may be implemented on other servers such as an IBM LAN. It is very similar to SMTP in that both are well documented, open, and use a text based envelope, heading, and message structure. Both are relatively easy to learn and both have much less complex structures than SNADS/DIA or X.400 protocols. This simplicity makes it quick to implement and reduces administrative tasks of the network. The reduced system overhead produces better message delivery times and facilitates the connection with different messaging systems.

Global MHS Message Structure

NetWare Global MHS defines and implements four items: the standard message format, the standard directory record format, the directory synchronization mechanism, and the message transport mechanism. The environment supports two messaging structure protocols: the old well-known structure, MHS 1.5, which uses the Standard Message Format - 70 (SMF-70), and the new structure, NetWare Global MHS, which uses the Standard Message Format - 71 (SMF-71).

SMF-70 and SMF-71 define the message structure for the Novell messaging system, which consists of three components: envelope, header fields, and an optional body. Attachments are supported but their structure is not defined in the standard.

The underlying Novell network file copying program automatically transports messages between users by copying them from disk directory to disk directory. These messages may be between local users or remote users on other servers.

Global MHS Addressing

Addressing in Global MHS is similar to the Internet SMTP Domain Name Service (DNS) addressing discussed in Chapter 11, "SMTP E-mail Services." The address consists of two general parts separated by the "@" (at sign symbol) with the left part being the most specific, and the right the more general. Like SMTP addressing, any subparts to either part are separated by a period. The following is the general format for an address in SMF-71:

user_name@workgroup

SMF-71 supports two address formats: basic and extended. The basic format is used mainly between Global MHS systems, while the extended format is used between Global MHS systems and outside delivery systems such as postal delivery, X.400, SMTP, SNADS, telex, and fax. The extended format lengthens the address by adding groups on the right side to support the addressing required by these outside delivery systems. This is an example format for the extended address form:

user_name@workgroup (comments) {exterior address} [server]

Both formats usually require the Electronic Messaging Association (EMA) printable character set. A message directed to user Ben Smith on an X.400 system may look like the following:

```
Ben Smith@Sales.Ace_Tool {X400:C=us;A=MCImail;P=ABC;S=smith;G=ben} X400GATE
```

Note how the right side of the address is a normal X.400 address string. See Chapter 12, "X.400 Interpersonal Messaging," for more detail on X.400 addressing.

MHS Message Structure

The structure of the message has three parts: the envelope, composed of keywords and values; the message heading, composed of keywords and values; and the message body, composed of optional text or formatted text. Optional text is unstructured and entered by the user. Formatted text is structured and contains the output of applications such as calendaring, word processing, and spreadsheet programs. Attachments are also supported. Figure 10-1 shows an example message formatted in SMF.

```
SMF-71
Send-to: Tim@Finance.ABCTool
Sender: Sally@Purchasing.ABCTool
Date-posted: 5-May-95 14:03
Attachment: $3456135
Attachment-name: Purchasing Log Apr95
MHS-id: 12345ADO9938473
To: Tim@Finance.ABCTool
From: Sally@Purchasing.ABCTool
Subject: Attached Purchasing Report
Message-id: 1234485939095948


Tim, please see attached Purchasing Report for April,
1995


                    Best Regards,
                    Sally
```

Figure 10-1.   *Example of an SMF-formatted message*

Message Envelope

The envelope in NetWare Global MHS is used by the transport mechanism to convey and switch the e-mail messages between users on the local server, on remote servers, and on other transport mechanisms. The envelope created by the application always starts with the SMF-71 (the older SMF-70 version, though valid, is not discussed here) in the first character position on the first line of the file. Envelope and header fields have unique names and may be intermixed. The body starts after the first blank line is encountered. See Figure 10-1 for an example of an SMF-formatted envelope.

Envelope fields used by message transfer agents (MTAs) to convey and deliver messages are listed in Table 10-1. The type of fields and the information they convey indicate the richness and depth of the message architecture. Envelope fields request services from messaging systems, such as to list the number of hops the message made to arrive at the destination or to expand a distribution list. The types of information found is these fields are fairly common and are also generally found in the X.400 and SMTP standards. For more information on the message structures of those systems,

Envelope Field	Field Content
Addresses-referred-to	This is the address for the nondelivery or error report message that caused this message to be generated. This field is used in error and nondelivery messages.
Attachment	This is the name of the attachment file in the PARCEL directory. It is not the name of the actual submitted attached file.
Attachment-date	This is the last modified date for the attachments.
Attachment-encoding	This field states the character set with which the attachment was encoded.
Attachment-type	This field indicates the type of file, such as a Microsoft Word or Excel file
Attachment-name	This is the actual name of the attached file as submitted to the messaging API.
Date-delivered	This field contains the date and time the message was delivered to the end application.
Date-posted	This field states the date and time the message was submitted to the MTA.
Date-transferred	This is the date and time the message was transferred between MTAs. There would be one entry for each MTA-to-MTA hop.

Table 10-1. *Global MHS Envelope Fields*

Envelope Field	Field Content
Delivered-to	This is the address to which the message was delivered. It is used in delivery and error reports back to the source user.
Destination-application	This field contains the name of the application for which the message is bound.
DL-expansion	This field holds information on the Distribution List expansion history.
Error-report	This is the type of error encountered.
Expiration-date	This field contains the date and time after which the message is no longer valid.
Header-encoding	This field describes the character set used to encode the fields in the heading.
Hop-count	This field is the number of MTAs that have handled the message.
MCB-options	This is used to request services from the MTA system. Nine types of services and meaningful combinations may be specified: Return of contents Receipt notification Nondelivery notification Grade of delivery Alternate application delivery allowed or disallowed Delivery notification, designate-delivery Designate delivery Nonreceipt notification Document translation allowed or prohibited
MCB-type	This field identifies the type of message: message, delivery or nondelivery report, and receipt or nonreceipt notification.
Message-encoding	This is the character set used to encode the message text body.
MHS-id	This is the MTA identification number.
Notification-correlator	A unique application identifier is used to match messages.
OEMappname-proprietary_keyword	A proprietary keyword is listed for the application name.
Received-by	This field is used to return more detailed information about the delivery in report messages.

Table 10-1. *Global MHS Envelope Fields* (continued)

Envelope Field	Field Content
Refused-by	This is used for returning more detailed information in delivery report messages.
Sent-to	These are destination addresses used by the MTAs. The system places messages in this field that came from the TO:, CC:, and BCC: fields.
Sender	This is the originator's address to which error and notification report messages will be returned.
Session-id	This is an MTA data field.
Signature	Indicates the version of the protocol supported, such as SMF-70 or SMF-71.
SMF-version	This field shows the SMF version as either 70 or 71.
Via-host	This is the server where the message was created.
X-Any_text_string	This is an experimental field much like those in RFC 822 headings. Just about anything may be included in this field.

Table 10-1. *Global MHS Envelope Fields* (continued)

refer to the section "X.400 and Internet Message Mapping" in Chapter 7, "Strategies for Connecting Disparate E-mail Systems."

Message Heading

The fields in the SMF heading are similar to those in the SMTP/RFC 822 and X.400 headings. (See Chapters 11 and 12.) The heading and envelope fields of Internet SMTP/RFC 822 and SMF are mingled, while in the case of X.400, they are separated into distinct areas. The user submits a file in the SMF-71 message format and the message system reformats the envelope fields, deletes some submitted fields, and creates a file for transmission. At the end of the first blank line, the envelope and heading fields end. A blank line is symbolized by <CR><LF>. After the first blank line, the body of the message starts. This is the same technique used in the SMTP/RFC 822 protocol.

The heading fields of the SMF and SMTP envelope and heading are user readable whereas the IBM DIA and the X.400 are binary encoded—not plain text. Some applications pass the envelope fields to the user. The heading fields listed in Table 10-2 are those used for Interpersonal Messaging and only have meaning to the end applications. They are not used by the MTA, but are used by the end-to-end applications.

Heading Field	Field Content
Application-name	This field contains the application of the originating user.
Authors	When the author of the message is not the sender of the message, this field is used.
BCC	This signifies the address of each recipient who receives a blind copy of the message.
Comments	This field is reserved for comments.
Conversion-id	This is used to reference a message series with the same topic (a message thread in the SMTP world).
Copies-to	This field lists the addresses of users receiving copies.
Date	This is the date the message was created in dd-mmm-yyyy hh:mm:ss -hhmm format
Date-received	This is the date the message was received.
Form-type	This is the type of form in a form structured message.
From	This signifies the name and address of the originator of the message.
Importance	The sender rates the importance of the message: High, Normal and Low.
In-reply-to	This field refers to the original message in a replying message.
Keywords	These are the keywords used by the application to classify the message.
Message-id	The unique message identification used to reference the message is placed in this field.
Message-type	The type of message is specified in this field.
Original-copies-to	This field records exploded DL names.
Original-to	This is the way the To: field looked before it was exploded into Distribution Lists.
Originator	The original From: field in a series of forwarded messages and replies are placed in this field.
Reply-copies-to	The addresses of those who should receive copies of the reply are placed in this field.
Reply-to	Those to send copies to for replies are in this field.
Re-sent-by	This is the field in a forwarded message that identifies the last forwarder of the message.
Respond-by	An answer is expected by this date.

Table 10-2. *Global MHS Heading Fields*

Heading Field	Field Content
Ret-message	This field identifies the message for which this is a report.
Sensitivity	This field contains the rating given by the originator of the message's sensitivity. Levels are Normal, Private, Personal and Company-Confidential.
Subject	The topic of the message is inserted in this field.
Summary	In this field, a concise presentation of the substance of the message contents is included.
To	The address(es) of the primary recipient(s) are placed in this field.
X-Any_text_string	This is an experimental text string.

Table 10-2. *Global MHS Heading Fields* (continued)

Message Body

The body of the Global MHS message may be free form or a structured text such as that used for forms or calendar applications. The body is optional and need not be present in the message. NetWare Global MHS supports several different types of services: person to person, person to/from program, and program to program. Not all of these conveyances need a message body.

Transport

NetWare Global MHS submits messages to the transport system by copying a properly formatted SMF-71 file to the directory named *mhs\mail\snd*. Attachments are copied to an adjacent directory called *mhs\mail\parcel* and pointed to by the "attachment" envelope field in the e-mail message. The transport mechanism copies the file and attachments to the appropriate user directory on the local or remote system. In some cases the file and attachments may be copied in a store-and-forward manner to intermediate servers before finally being copied to the destination user's mail directory.

Some transport servers convert different attachments from one format to another but this is not part of the base function of NetWare Global MHS. However, the architecture does support the "plug and play" of such converter modules from other vendors.

Detailed Processing Steps

The major steps used by the MTA in message transfers between NetWare Global MHS users are listed here. These steps assume a network file system where disks on remote network servers are mountable by other servers. See Figure 10-2 for a description of the file system.

1. The user creates a message in SMF-71 format. The message is placed in the outbound mail directory *mhs\mail\snd*. Any attachments to the message are placed in the *mhs\mail\parcel* directory. A field in the message envelope, "attachment," contains the name of the attachment file in the *mhs\mail\parcel* directory.

2. The transport program periodically checks the *mhs\mail\snd* directory for outbound mail, which it processes in one of two manners:

   a. If the addressed user is on the same server, the message is moved to the user's in-box message queue, *mhs\mail\user\ "use name"\mhs,* and the attachments, if any, are moved to the adjacent directory *mhs\mail\user\ 'username'\iparcel.* At this point the message and any attachments have been delivered to the user.

   b. If the user resides on a different server, the message is moved to the outbound queue directory called *mhs\mail\queues.* The MHS MTA program periodically moves messages and specified attachments to the remote system of the addressee and puts the message and attachments into the *mhs\mail\snd* and *mhs\mail\parcel* directories, respectively.
   Now the message is in the same state as any other local message and may be processed as in step 2a.

3. The message has been delivered to the addressee's in-box queue.

Disk Directory Structure

Instead of sending the e-mail to message queues, as is the case for the X.400 and SNADS systems, or connecting to the appropriate remote system for direct delivery, as in the case of SMTP, NetWare Global MHS copies messages to the directory where the transport programs copy the file to the appropriate account on the destination system. The message is delivered to the user by placing it in the inbound mail account of the destination user. See Figure 10-2 for a description.

Address Directory

You may choose which server will participate in the Global MHS directory for ongoing directory synchronization. The synchronization method updates each participating server through directory update messages. The synchronization updates e-mail users, interserver e-mail routes, and distribution lists. The synchronization

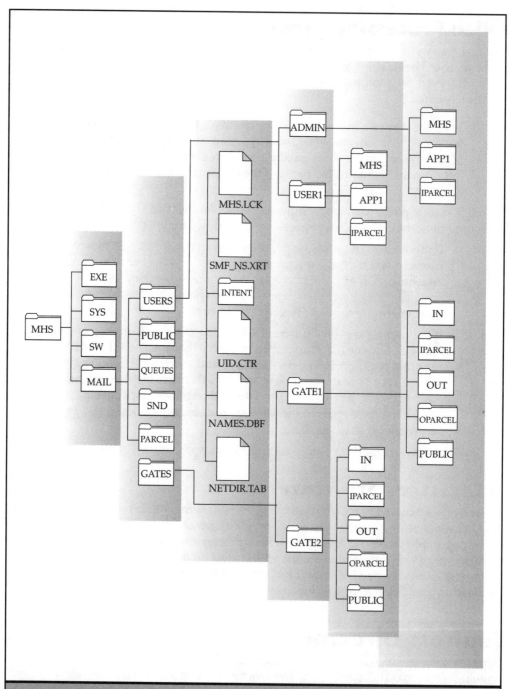

Figure 10-2. *Novell NetWare Global MHS Directory Structure*

Directory Field	Field Length
Fullname	66 bytes
SMF Name	256 bytes
Short SMF Name	18 bytes
Department	20 bytes
Title	20 bytes
Phone Number	28 bytes
Other information	104 bytes

Table 10-3.   *Global MHS Directory Fields*

process is elegant in its simplicity and speed. The disadvantage is that inaccurate information can also be replicated quickly through all the participating servers.

A user directory record is 512 bytes in length and holds substantial information. To compare the types of data contained in the record, see Chapters 6 and 8 for analogous information about X.500, Enterprise Address Book, and other products. The data in these records determines how easy it is to synchronize directories with other systems. The data has the fields listed in Table 10-3.

Summary

Novell NetWare Global MHS architecture is easy to learn and use for creating messaging applications. The message format is easily interpretable by the programmer and user, and easily accommodates forms and other applications. In many ways it resembles the Internet SMTP and RFC 822 message transfer and message structure standards. Of the four multiplatform protocols discussed in this section (the others being SNADS, SMTP, and X.400), Novell NetWare Global MHS is the only one that supports attachments separate from the actual message file. The others all include the attachment in the message file itself. The latter has benefits in message integrity. Overall this is a well integrated set of services.

Chapter Eleven

SMTP E-mail Services

The Simple Mail Transfer Protocol (SMTP) is the MTA-to-MTA protocol for the Internet, the largest e-mail network in the world, composed of millions of interconnected systems spanning over 50 countries. SMTP is used to deliver e-mail between different types of computer systems and environments. The Internet is a composite of many national networks, such as MILNET and NSFNet, in addition to a large number of commercial and educational institutions.

The Internet standards have no supporting directory structure, as is the case with SNADS and X.400. Because of this, the Internet community has shown a lot of interest in the X.500 directory standards, discussed in Chapter 8, "X.500 Directory Systems."

Since the Internet functions as a large virtual network, interaction with it must be flexible, uncomplicated, timely, and user friendly. The most common transport protocols that enable disparate computer systems to communicate are SMTP and X.400, and they differ considerably in their approach. This chapter addresses the naming and addressing conventions used in SMTP—in the message transport system and message structure.

SMTP Messaging Architecture

IBM's Systems Network Architecture Distribution Services (SNADS) and Novell's NetWare Global MHS standards are vendor owned, whereas the Internet and X.400 are international in nature and have a very large following. The SMTP transport protocol was defined in Internet RFC 821 in 1982. The SMTP message structure was defined in RFC 822 and subsequent RFCs.

SMTP Addressing

Early SMTP addresses stated the exact route the message went through to be delivered, and users had to remember complex machine names. The Domain Name Service (DNS) addressing structure was implemented later so users addressed domains and not physical machines. A *domain* is an entity, such as an organization, company, university, or government agency, managing its own connections to the Internet. When the MTA of a domain receives a message, it determines the detailed physical address needed to deliver it. The DNS contains routing information for all hosts within a specified network.

The DNS address has two parts separated by an @ sign. The part on the left of the @ sign is assigned by the local SMTP administrator, while the right side contains the name and type of domain. An example of an SMTP address is as follows:

RickJones@AOL.COM

RickJones is a user's mailbox name on America Online (AOL), and AOL is a domain name registered under the top level domain called COM. A *top level domain*

(TLD) is similar to the root of a hierarchical directory tree and is managed by an entity such as a government or military.

Users that are part of a recognized domain acquire SMTP addresses from their local domain administrator. Individual users can access the Internet through several value added networks (VANs) and online services such as CompuServe.

There are two types of TLDs: those signified by two-letter country codes defined in standard ISO-3166 and those signified by generic codes. Generic TLDs contain a collection of similar organizations or domains that can be grouped into one of seven categories. Five are international; the other two are used only in the United States:

Generic TLDs	Usage
COM	This domain is used to register commercial entities such as corporations and companies.
EDU	This domain is used to register educational entities. The current preference is to register only universities and four-year colleges under this name.
NET	This domain is used to register computer and network providers such as MCI, IBM, AT&T, and Sprint.
ORG	This domain is intended for miscellaneous registrations.
INT	This domain is for registering organizations established by international treaties.
GOV	This domain is used only by the United States government.
MIL	This domain is used only by the United States military.

The organization managing each specific TLD is responsible for assigning the subdomain names. This may be done differently for each country and each generic TLD. Since addresses are hierarchical in structure, they frequently consist of several subdomain names in addition to the TLD. For instance, the domain name *eng.ucf.edu* would represent the Engineering Department (eng) of the University of Central Florida (ucf) within the TLD for educational institutions (edu).

An example of a country TLD would be the United States domain, US. The structure of the subdomains in the United States is a geographical governmental hierarchy, such as:

local_part@entity_name.locality.state-code.US

For example:

TedSmith@ABCTOOL.Dallas.TX.US

This is a valid address under the United States top level domain, US.

Transport

When an SMTP e-mail message is sent to an SMTP user on another system, the originating system connects directly to the destination system using the specified address and, through interprocess communications, exchanges a number of half-duplex data blocks to transfer the message to the end user. The exchange starts when the originating system passes the addresses to the destination system one at a time. The destination system responds as to whether it will accept the message for that address. When the message must be routed through a large network, the address of the message may be for an intermediary system instead of an end user, in which case the intermediate machine passes the e-mail to yet another intermediary system or to the specified destination system.

The user sends a message that looks like the one shown in Figure 11-1. The end-user system connects to a system named endnode2 and commences the following exchange, as detailed in Figure 11-2. Note that a period following the text of the message, "fjsddjdfj," symbolizes a blank line and indicates the end of the message body. If the originating node (endnode1) finds a period that starts a line in the actual message body, another period is temporarily attached to that line by the originating node. The destination system then removes the first period from the text body before passing it on.

Information is transferred one line at a time and each line cannot be over 1,000 characters long. The end of a line is indicated with a <CR><LF> (carriage return line feed). To ensure that the message is transportable across all Internet-connected systems, a message cannot be over 64K in size, roughly 25 plain text pages.

The HELO command, with its ensuing parameter list, tells the other system that the exchange uses the SMTP protocol and that the message is from endnode1. The MAIL command notifies the other system that an SMTP e-mail message is coming. If the response to the MAIL command is OK, then the sending system follows with the recipient's RCPT (receipt) command. If not, the connection is terminated. The recipient command RCPT contains one of the message addressees. In Figure 11-2, the recipient is Rik@endnode2.com.

If the receiving machine, endnode2, has an e-mail user named Rik, it will accept the address by returning an OK. The data of the message is then sent by endnode1,

```
Date: Jan 1, 1999
To: Rik@endnode2.com
From: Sally@endnode1.com

This is a test message. This is a test message.
This is a test message. This is a test message.
```

Figure 11-1. *A simple SMTP message*

```
        ENDNODE1                    ENDNODE2
                    start session

                            <---220 ENDNODE2.COM smtp ready

    HELO ENDNODE1.COM---->

    MAIL ENDNODE1.COM---->
                            <---250 OK
    RCPT RIK@ENDNODE2.COM------>
                            <---250 OK
    DATA----->
                            <---354 start mail transfer


    fjsddjdfj---->


    .
    QUIT----->
                            <---250 OK

                            <---221 ENDNODE2.COM

                    end session
```

Figure 11-2. *An SMTP message exchange*

specifying the DATA command, and endnode2 replys with an OK. The data section is completed when endnode1 transmits a <CR><LF>.<CR><LF> (carriage return, linefeed, period, carriage return, linefeed) sequence. If the data transmission is acceptable, endnode2 responds with a 250 OK. If the sender is finished sending messages, endnode1 issues a QUIT. The receiver will then issue the command 221 endnode2.com, ending the session.

> **NOTE:** *With the advent of DNS, Internet addresses can be a logical alias instead of the actual physical address. Therefore, rather than giving the actual endnode2.com string, which specifies the address name, the address could be rik@companya.com. The originating system looks in the DNS and finds that Company A users reside on the machine named endnode2. This is sufficient information to enable the originating system to send the e-mail message. The originating system then would have "Rik," the user name; "endnode2," the system serving the user; and "com" (commercial), the major network component type.*

The transport envelope of an SMTP/MIME message contains heading and envelope fields, as well as MIME multiple body parts, as shown in Figure 11-3.

Other commands besides those listed previously are also available in the SMTP exchange. A few examples follow:

Command	Function
RESET	Aborts the current e-mail transaction. The receiving system must respond with an OK.
VRFY (Verify)	Asks the receiving system to verify that the user address string is correct and known.
EXPN (Expand)	Asks the receiving systems to verify that the address string is a distribution list and return the members of the list.
HELP	Displays help function.
QUIT	Send an OK from the recipient and then close the communications channel.
TURN	Asks the receiver to turn the communications channel around, whereby the receiver assumes the role of the initiator. This is useful if system A contacts system B to send mail and system B needs to send mail to system A before the channel is terminated.

Extended SMTP

E-mail users are sending increasingly larger messages than ever before. Networks designed to support a small maximum message size with a minimal amount of overhead are now laboring under this heavy traffic. To enable users to send larger messages and keep the overhead at an acceptable level, Extended SMTP was introduced in 1993.

RFC 1425, RFC 1426, and RFC 1427 are the documents that extend the SMTP protocol by registering new commands, service extensions, and command parameters.

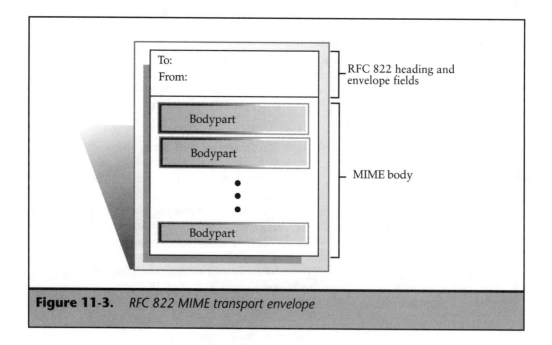

Figure 11-3. *RFC 822 MIME transport envelope*

One of the most important service extensions to SMTP is the support for full eight-bit data transmittals. Another is the introduction of the keyword "SIZE," which lets the destination SMTP node specify the maximum message size it will support before the exchange takes place. The basic SMTP supports eight-bit bytes with the high-order bit always zero. This eight-bit SMTP extension enhances functionality and reduces transmission overhead by about 25 percent. It makes SMTP a viable protocol for the future.

Message Structure

The SMTP has two types of message structures. The original RFC 822 was designed for ASCII message data only and was ideal for the uncomplicated nature of message exchange on the Internet. With the increase in message traffic, message size, and requirements for end users to exchange complex messages containing more body parts, the Multipurpose Internet Mail Extension (MIME) was developed via the RFC process (RFC 1341) by Borenstein and Freed.

RFC 822 Message Structure

Table 11-1 illustrates the envelope and heading fields used in RFC 822 messages. Each of these are in ASCII and do not require decoding by the computer to be read by the receiver. These fields are very similar in name and description to the X.400 envelope and heading fields discussed in Chapter 12, "X.400 Interpersonal Messaging."

MIME Message Structure

MIME provides the ability to encapsulate different content types within the body of the message. Before the initial June 1992 MIME specification RFC 1351, the Internet was only able to transmit and receive ASCII-type data. If you wanted to send binary data, you had to first convert it to ASCII (see note below). The most common method of conversion up until now has been the UUENCODE/UUDECODE facility, although MIME is quickly replacing it. The latest version of MIME, specified in RFC 1521 and approved in September, 1993, added the ability of Internet e-mail to handle binary and text data, as well as multiple body parts, without conversion to ASCII.

NOTE: Before sending a graphic such as a drawing or picture without MIME, you must first run the graphic file through a UUENCODE package and put the resultant ASCII-type output into the e-mail message. The recipient processes the file by running UUDECODE to produce the original binary code for the graphic. The Internet ensures that only messages less than 64K are transported through all gateways and systems. If the graphic is larger than 64K, then UUENCODE breaks it into 64K ASCII-encoded segments to be sent in multiple e-mail messages. When all the parts are received, the UUDECODE reconstructs the original file from all the parts and decodes the message.

RFC 822 Message Field	Description
Auto-forward	Address of the account that autoforwards the message.
BCC	Blind copy addressee(s). These are not known to the To or CC addressees.
CC	Carbon copy addressee(s).
Comments	General remarks.
Date	Date the message is sent.
Discarded-X400-IPMS-Extensions	Lost components after converting an X.400 message to an SMTP/RFC 822 message.
Encrypted	Indication that the message is encrypted.
Expiry-date	Date the message is no longer valid.
From	Name and address of the originator of the message.
Importance	Importance of the rating given by the message sender. Choices are Urgent and NonUrgent.
In Reply To	Message ID of the original message that this new message is in in reply to.
Languages	Language in which the message is written.
Message-Id	Unique message identifier.
Mime-Version	Indicates MIME body part type.
Obsoletes	Message ID of message(s) no longer in use.
References	Other message(s) to which this one refers.
Reply-by	Date by which a reply should be received.
Reply-to	Addressee(s) that should be replied to when using the reply function.
Resent	Address of the person who resent the message.
Sender	Sender of the message. This may be different from the "From:" entry.
Sensitivity	Sensitivity rating given by the message sender. Choices are Personal, Confidential, or Private.
Subject	Topic of the message.
To	Address(es) of receiver(s).

Table 11-1.   *RFC 822 Message Envelope and Heading Fields*

MIME Body Parts

MIME provides a multiple body part architecture used by SMTP to transport binary and text code within the same message. It does not get around the 64K limit, which is a transport limit, based on total size of the message. Messages larger in size than 64K are transmitted in separate 64K segments, which the system then reassembles.

Boundary strings are used to separate body parts. These strings are unique over the entire message and are defined on a message-by-message basis. Each is on a line by itself, starting with "- -" (two hyphens) and ending with end-of-line characters like a carriage return. The heading field defining the boundary string for this message might look like this:

```
boundary=--jfjd9dk38dmfmf3840fm
```

This string of characters would not appear anywhere in the bodies of the messages. Within the message body itself, the boundary string would appear as follows, separating a simple text message from a file:

```
Hi, Henry. This is your sales report.
Please reply. Thanks, Ted
--jfjd9dk38dmfmf3840fm

Quarterly Sales Report...
--jfjd9dk38dmfmf3840fm--
```

MIME specifications currently support seven body part types: text, multipart, application, message, image, audio, and video. Support for the various types, which are described next, enables the Internet to carry larger and more complex messages in preparation for the advent of multimedia messaging.

TEXT The text body part enables a message to contain simple message data such as ASCII, and can be transported using the current seven-bit ASCII strings used on Internet. This is the most rudimentary form of message content specified within MIME. It may also contain other subtype text strings known as *plain,* or unembellished text, and richtext. The *richtext* subtype is used to handle simple text format protocols that support boldfacing, italicizing, indenting, and so on. The richtext protocols are a greatly reduced subset of the SGML commands. Standard Generalized Markup Language (SGML) is an international standard that separates a document into discrete, structured parts so that the document can be revised in different computer environments. Currently, two categories of character sets are supported: Charset= US-ASCII and ISO-8859-1 through ISO-8859-9.

MULTIPART The multipart body part consists of several body parts containing unrelated data. MIME permits the user to break the content down into subtypes. The four initial subtypes are mixed, alternative, parallel, and digest. Only seven-bit, eight-bit, or binary may be used for multipart content type encoding. If other encoding is required, each body part will specify that encoding in its header.

■ *Mixed* The mixed multiple body part subtype is the most frequently used; it ensures that a number of very different message content types, such as text, graphics, or images, can be transmitted in the same message.

- *Alternative*   This subtype presents the same data in different formats, such as a word processing document in three representations such as ASCII, Word for Windows Version 2, and WordPerfect 6.01.

- *Parallel*   This subtype contains body parts that must be viewed at the same time. This type is useful when documents are linked with a utility such as Hypertext.

- *Digest*   This subtype is used when all the body parts are messages in their own right. It is very important that an e-mail gateway interpret that the message body is a nested message, as opposed to a video image or graphic, because the gateway has to handle each body part differently.

MESSAGE   A message body part contains other messages, like forwarded or transferred messages. It is the most basic body part in MIME, and its subtypes are

- *RFC 822*   This subtype is the specification for a complete standard Internet e-mail message and is the primary and most frequently used subtype for this content type.

- *Partial*   This subtype allows messages to be sent in parts through the e-mail network. This is necessary when the message has exceeded the 64K limit imposed by the Internet transport system.

- *External-Body*   This subtype is for specifying larger data files, such as text, video, audio, or other that are not contained within the message. The body parts reference the location of the data on mail servers or anonymous FTP servers. *File Transfer Protocol (FTP) servers* permit any user to locate and download a file to their own computer.

IMAGE   The image body part contains *time varying images,* or images that contain movement-like motion pictures and full motion video. The current subtypes are

- *MPEG*   Motion Picture Experts Group (MPEG) is the standard for digitally compressing movies.

- *GIF*   CompuServe's Graphic Image Format.

AUDIO   The audio body part contains sound data such as voice or music. The basic subtype indicates eight-bit, Integrated Services Digital Network (ISDN), or u-law encoding, with a sample rate of 8,000 Hz.

APPLICATION   Application body parts, generated from computer application programs, contain spreadsheets, calendar information, word processing documents, presentations formats like WordPerfect or Microsoft Word. Its current subtypes are

- *Octet-Stream*   This subtype is used for binary data that does not need or have an interpreter.

■ *ODA*   This subtype is the Office Document Architecture as defined by the International Telecommunications Union (ITU).

■ *Postscript*   This subtype is defined by Adobe Systems, Inc., and supports high quality postscript printer output. This subtype should not be used with nonprinter interpreters, because the information contained in a postscript file is so rich that sending this format may give the receiver information about the sender's access to files and other rights.

All the preceding body part content types can be defined in a MIME envelope. These are being used exclusively on the Internet at the moment, though there is some discussion about using them on other transports like X.400.

MIME, like IBM DIA and X.400 P2, is recursive in nature. A MIME body part message can contain a nested MIME format e-mail message containing a MIME-format e-mail message, containing a MIME-format e-mail message, and so on.

For example, an RFC 822 message contains a MIME format of content type of message. The subtype is RFC 822. RFC 822 means the message contained within is possibly a MIME formatted message that has a content type message with a subtype of RFC 822. The protocol specification will allow recursion to infinity, even though the implementation will most likely stop at a much lower value.

Figure 11-4 shows the logical structure of a MIME message.

MIME Data Encoding Techniques

The current SMTP network only supports seven-bit ASCII, up to 1,000 characters per line of data, and a normal message length of 64K. Longer messages are possible after being segmented into manageable parts, but the maximum length that will go through any gateway is still 64K. Enhanced SMTP, which supports binary data exchange, is currently being implemented. However, until it is universal, seven-bit encoding is still necessary for most eight-bit messages. The RFC 821 (SMTP) compliant networks will not handle binary data contained in the MIME structure. Users have to use UUENCODE/UUDECODE to convert their messages until all networks are converted to the enhanced SMTP version.

The RFC 1521 specifies that the body of the message can be encoded in a form that will be transportable by the SMTP network. To show this, a new field, called *Content-Transfer-Encoding*, has been added to the header of the RFC 822 message. It may have one of the following six different encoding values:

BASE64   Base64 is for any series of octets and is used in Privacy Enhanced Messaging (PEM), specified in RFC 1113. Binary input strings are converted to a series of 65 ASCII characters which are the only ones that are represented the same in ISO 646, US ASCII and EBCDIC. This encoding takes a series of three octets and outputs four ASCII characters to represent them. The resulting data can go through all RFC 821 gateways, and the end message body is much smaller in size than the starting message body.

```
Sender: ifip-emailmgt-request@ics.uci.edu
Received: from ics.uci.edu by arl-img-2.compuserve.com
(8.6.4/5.930129sam)
    id PAA06676; Wed, 19 Jan 1994 15:59:24-0500
Received: from ics.uci.edu by q2.ics.uci.edu id aa 16108;
    19 Jan 94 12:46 PST
Received: from zephyr.isi.edu by CNRI.Reston.VA.US id aa17031;
    11 Jan 94 15:51 EST
Received: from akamai.isi.edu by zephyr.isi.edu
    (5.65c/5.61+local-16)
    id<AA25678>;tues, 11 Jan 1994 12:51:26-0800
Message-Id: <199401112051.AA25678@zephyr.isi.edu>
To: IETF-Announce:;@compuserve.com
Subject: RFC 1566 on Mail Monitoring MIB
Cc: jkrey@isi.edu
Mime-Version: 1.0
Content-Type: Multipart/Mixed; Boundary=NextPart
Date: Tue, 11 Jan 94 12:51:19 PST
Sender: ietf-announce-request@ietf.cnri.reston.va.us
From: "Joyce K. Reynolds" <jkrey@isi.edu>

--NextPart

A new Request for Comments is now available in online RFC
    libraries.. ...

--NextPart
Content-Type: text/plain
Content-ID: <940111124659.RFC@ISI.EDU>

SEND /rfc/rfc1566.txt

--NextPart--
```

Figure 11-4.   *A MIME multipart message*

EIGHT-BIT   Eight-bit means that lines are of the same form as they are in seven-bit encoding. However, the high order bit of the byte, the one not used in seven-bit ASCII, may be set in some of the characters. Eight-bit also means that the body has not been encoded.

BINARY   Binary means that there is not a line length limit within the message. It also means that the body has not been encoded.

QUOTED PRINTABLE ENCODING   This encoding value is for data that generally uses an ASCII character set. Instructions on how to encode this type are contained in

RFC 1521. A key objective of this encoding is that the end result is readable without conversion for the encoded type. It allows unsophicated message user agents to convey data, the format of which may be a little off, but which is readable by the end user.

SEVEN-BIT   Seven-bit is the default value when the Content-Transfer-Encoding header field is not present in the header. This means that the data is of the type specified in the RFC 821, seven-bit US ASCII code and has not been encoded.

X-TOKEN   This value is for defining a nonstandard encoding which has been put in place by mutual agreement between the parties to the transfer.

Address Directory

The SMTP architecture does not define an address directory. Users find names by enrolling in Distribution Lists, using a common utility program called FINGER to search on their system, and a query facility called WHOIS to find addresses. The Internet community has a very large X.500 directory in production as part of the Paradise pilot project.

Summary

Due to its worldwide acceptance, critical mass of end users, and user-friendly functionality, the Internet, in all likelihood, will become the technological infrastructure for the national information superhighway. The advent of MIME means the Internet is well positioned to carry multimedia messages containing such heretofore unthinkable content types as movies, animated graphics, holograms, encyclopedic volumes of information, and so on. These multimedia messages will be exchanged using SMTP.

Chapter Twelve

X.400 Interpersonal Messaging

The Consultative Committee for International Telephony and Telegraphy (CCITT) realized that the electronic community needed a way to transfer information between different computer systems and through different vendors operating throughout the world. The standard developers also knew that each individual, computer application, and network device must have a unique and unambiguous name in order to be addressable for message transfer. In 1984, they released the first set of X.400 recommendations that became the most widely accepted standard for message handling. The X.400 standard provides the model for designing standardized software that enables the global exchange of messages in a store-and- forward environment without regard to the user's e-mail or computer system. The standard describes message handling system architecture, the means by which messages are transferred between cooperating open systems, the message structure, the service elements, and addressing.

There are four types of messages supported by the X.400 standard: Undefined, 1984 Interpersonal Messaging (IPM), 1988 Interpersonal Messaging (IPM), and EDI:

- *P0* describes the Undefined body type
- *P2* describes the 1984 IPM body type
- *P22* describes the 1988 IPM body type
- *P35* describes the 1992 EDI body type

The Undefined message type does not have a heading and body structure. The other three have a separate heading and body that is defined in the standard.

The discussion on X.400 in this book is limited to the X.400 Interpersonal Messaging (IPM) 1988 P22 body type, which is the actual X.400 e-mail message. P22 is often referred to as the 1988 IPM P2, and P2 as the 1984 IPM P2. Since we are discussing the architecture and broad concepts of the 1988 X.400 standards, there should be no confusion between the references to P2 for the Interpersonal message of the 1988 standard.

X.400 Messaging Architecture

X.400 Message Handling System (MHS) Interpersonal Messaging (IPM) is composed of three operational components with four transport protocols used to communicate, in a peer-to-peer manner, among the components. The protocols are P1, P2, P3, and P7. These protocols specify the rules and behaviors for interactions between the components. Figure 12-1 depicts the major components of the X.400 messaging handling environment. A definition of each of the operational components follows.

MESSAGE USER AGENT (MUA)   The MUA is the interface between the end user and the e-mail system. Examples of MUAs are the graphical main menu and subsequent application screens in ALL-IN-1, cc:Mail, and QuickMail with which the user interacts to perform messaging related tasks. Whether an MUA supports X.400 or other protocols generally is not easily discernible to the end user. The MUA communicates with the other portions of the messaging system using one of three protocols:

- *P7* communicates with the message store (MS) or remote file cabinet.
- *P3* communicates directly with the message transport agent (MTA) by either the MUA or the MS.
- *P1* is used for MTA-to-MTA transfers.

MESSAGE STORE (MS)   The MS communicates with the MTAs for message submission and retrieval and stores messages on behalf of the user. The MS is much like a single user post office (PO) in which messages are stored for just one user. Each user has at least one MS, but does not have to use it; the user can communicate directly with the MTA. The MS communicates with the user through the user's MUA using the P7 protocol, and communicates with the MTA using the P3 protocol.

MESSAGE TRANSFER AGENT (MTA)   The MTA is the X.400 system entity that transfers messages between users. MTAs can also convert documents and expand distribution lists within messages, adding the distribution list addresses to the various recipient fields in the message.

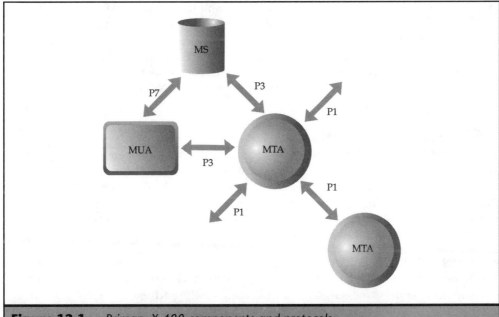

Figure 12-1. *Primary X.400 components and protocols*

X.400 Addressing

Intersystem, interproduct, and intercompany addressing is a major issue in using messaging systems. One of the major obstacles to the expansion of e-mail use is the difficulty of using intersystem addressing, especially in X.400 systems. Internet addressing is easier to use and is more user friendly.

The Electronic Messaging Association (EMA), based in Arlington, Virginia, is an association of electronic message providers and users with a common interest in all aspects of the electronic messaging industry. All of the large commercial electronic message carriers, most e-mail vendors, and many Fortune 500 companies are members. The EMA convenes regularly to discuss such issues as e-mail system management, X.400 addressing, system interoperability, government initiatives, X.500 directories, and wireless e-mail.

An ongoing project of the EMA has been an "X.400 Reference Guide" written for e-mail administrators that deals with the problems associated with addressing messages between the private and the public domain. It explains how to address users on value added networks (VANs) such as MCIMail, ATTMail, Mark400, IBMX400, Telemail, and other VANs using the X.400 addressing methodology. This guide is constantly updated, and the next release will include Internet addressing and connectivity with various service providers.

X.400 Address Methodology

Attributing a globally unique, easily understood, and totally unambiguous e-mail address to an individual, a computer application, or a device has been a daunting task for those who have adopted the X.400 standard. As more and more organizations, service providers, and vendors have implemented X.400 message handling systems, variations in addressing conventions have grown exponentially. Unlike the addressing convention on a standard U.S. Postal Service envelope, the X.400 standard permits multiple variations of electronic addresses.

X.400 Address Forms

The X.400 address string is unusual. Unlike most addresses which are a sequence of characters, alphabetic or numeric, defining the user's e-mail address, X.400 allows one of four methods to be used. The most widely used form of X.400 addressing is the Mnemonic O/R (Originator/Recipient) Address—a form most users feel is unfriendly, hard to use, and holds back the spread of fully networked e-mail between different systems. Descriptions of the four X.400 address forms follow.

MNEMONIC O/R   This addressing method is defined by a sequence of parameters. It is the best known and the most widely used method. Designed to be user friendly and help users address messages without looking up the address in the directory, it is nevertheless difficult to understand, remember, and use.

TERMINAL O/R This method provides a way to address terminals on different networks. Some older e-mail systems do not have mailboxes, and mail is delivered to the user terminal address.

POSTAL O/R This method is used to specify a physical (paper) mail delivery through the postal service.

NUMERIC O/R This form provides a means for identifying a user via a series of numbers, like a phone number. The series of numbers represents the user's address.

Form 1, Variant 1 Mnemonic X.400 Addressing

The first entry on the preceding list is also called Form 1, Variant 1, and is the most common electronic messaging X.400 addressing method. The addressing scheme is hierarchical, the top level being the country and the bottom level, the individual's name.

The Form 1, Variant 1 parameter identifiers are defined in Table 12-1. These names and abbreviations, plus the values entered by the naming authority, are used to build the X.400 address. Completion of the required fields enables an end user, device, or application to be uniquely identified to the X.400 messaging community.

Name	Usual Abbreviation
Country	C
ADMD	A
PRMD	P
Organization	O
Organization Unit 1	OU1
Organization Unit 2	OU2
Organization Unit 3	OU3
Organization Unit 4	OU4
Surname	S
Given name	G
Initials	I
Generation qualifiers	Q
Domain Defined Attribute 1	DDA1
Domain Defined Attribute 2	DDA2
Domain Defined Attribute 3	DDA3
Domain Defined Attribute 4	DDA4

Table 12-1. *X.400 Form 1, Variant 1 Parameter Identifiers*

An example of an X.400 Form 1, Variant 1 address is:

```
C=UK
A=GOLD 400
P=Ace Rental
S=Reed
G=Chris
DDA1:EMS=Chris%Store21
```

The United Kingdom (*UK*) is the country through which the commercial intercompany carrier, *GOLD 400,* is accessed. *Ace Rental* is the private e-mail system that uses Gold 400. *Chris Reed* is an e-mail user who's e-mail address is part of the Ace Rental system. The *DDA1* field contains a type and value entry that is of use to the destination e-mail system. (Domain Defined Attribute (DDA) fields are used to help X.400 address other non-X.400 systems. Companies also use this field for tie-breaking information to ensure that each user's entry is unambiguous and unique.)

Following are descriptions of each parameter.

COUNTRY   The *country* is defined by a set of two-character codes. Examples are US (the United States of America), CA (Canada), and UK (United Kingdom).

ADMD   *Administrative Management Domain (ADMD)* is a commercial carrier offering services to subscribers. Examples are ATTMAIL, TELEMAIL, MARK400, MCIMail, IBMX400 and GOLD400. The length of the ADMD field can range from one to 16 printable characters.

PRMD   *Private Management Domain (PRMD)* is a private organization that manages its own messaging system. A name such as Boeing, Disney, or City of New York would appear in this field. The field length is one to 16 printable characters.

ORGANZIATION   This field usually represents an entity, company, division, or location, such as "Finance" or "Southwest."

OU   *Organization units* are for specifying organizational information such as company, division, function, location, and so on. OU1 is the broadest/highest level, and OU4 is the narrowest/lowest level. An X.400 address cannot contain information in OU2, 3, or 4 without first specifying information in OU1.

SURNAME   The *surname* is the user's last name. The field length is from one to 40 printable characters.

GIVEN NAME   The *given name* is the first name of the user. The field length is from one to 16 printable characters.

INITIALS   *Initials* are the first letters of all the user's names, in order from the first name up to but not including the surname. The field length is up to five characters. For example, the Initials parameter for Richard Henry Osborn Smith would be I=rho.

GENERATION QUALIFIER   The *generation qualifier* consists of short strings such as Jr., Sr., or III. The field length ranges from one to three characters.

DDAs   *Domain Defined Attributes (DDAs)* are used to specify information that is either specific to the destination system or that the organization uses to further identify a particular user, such as a mailbox number or employee badge number. Four DDAs are defined: DDA1, DDA2, DDA3, and DDA4. Information must be inserted in DDA1 before DDA2, 3, or 4 can be used. This field contains a type and a value, that can be separated by several delimiters such as a colon (:), an exclamation point (!), or an equal sign (=). The type and value entries in the DDA field are defined by the user.

There are several variations of these character strings. Also, while the rest of the fields contain only printable characters, DDA fields must support characters in the nonprintable string. The National Institute of Standards (NIST) has defined a way to encode these characters for the United States, which is covered later in the next section. The following are examples of DDA fields:

```
DDA1:TYPE=teletex
DDA1:TYPE=TELETEX
DDA1:EMS:PROFS
DDA1:RFC-822!KENB(a)I98.ISC.COM
```

You can see that the string following the delimiter character is case sensitive. That is, TELETEX may have a different meaning than teletex, Teletex, or teLETex. This string may also contain characters that are not supported by the X.400 convention. The field length is one to 128 characters.

A number of rules govern the construction of the X.400 address string. For example:

- No PRMD name may be used without an ADMD name.
- No given name may be used without a surname.
- Organization Unit 4 (OU4) cannot be used before OU3, OU2, and OU1 are used.
- Domain Defined Attribute 4 (DDA4) cannot be used before DDA1, DDA2, and DDA3 are used.

Form 1 Variant 1 Required Fields

The only parameters that must be present in the X.400 address are the country (C=), the Administrative Management Domain (A=), and one of the following: Private

Management Domain (P=), Personal name, Organization (O=), Organization Unit 1 (OU1=) or Domain Defined Attribute 1 (DDA1=). The personal name is composed of one or more of the following: surname (S=), given name (G=), initials (I=), and generational qualifier (Q=) and is also referred to as common name.

Table 12-2 and the following example show the encoding proposed by NIST to let the user input a superset of characters defined for the X.400 standard in DDA fields only. DDA fields contain the address of a non-X.400 system that uses a non-X.400 character set for addressing. For example, to encode this string

DDA1:SYS1=Chris%Store21

which uses characters that are illegal in the X.400 address string, you would input the following:

DDA1:SYS1=Chris(p)Store21

When the destination PRMD receives the message and transfers it to the gateway for DDA1:SYS1, the gateway reads the "Chris(p)Store21" string, converts it to Chris%Store21, and routes the message to Chris%Store21. The NIST encoding mappings are listed in Table 12-2.

Business Card Example

X.400 addresses now appear more frequently on business cards along with a user's Internet address, fax number, and so on, as shown in Figure 12-2. Address fields can be separated by commas, slashes, or semicolons. The fields listed may not be all that comprise the user's address. For example, since the X.400 address can be very long, DDA and initial fields may be left out, resulting in an inoperable address.

Intersystem Addressing Issues

Addressing between different systems is a complex issue because of the different addressing conventions, addressing character sets, and naming conventions. These are fairly ADMD-specific, so one learns how to send e-mail from their source ADMD to users on other ADMDs; however, you can't always just copy the address string off a business card and make it work. It may have to be adjusted to the source system format. This is often a problem for novice users.

Several ADMDs use different conventions that must be specified in an X.400 address. These conventions are closer across ADMDs than they were several years ago; however, there is still enough difference that users can become confused. The following X.400 address is shown in the business card example in Figure 12-2:

C=us; A=Telemail; P=Smidley; OU1=HQ; S=Smith; G=Ben

Character	Description	Encoded Representation
- %	Percent sign	(p)
- @	At sign	(a)
- !	Exclamation point	(b)
- "	Quotation mark	(q)
- _	Underscore	(u)
- (	Left parenthesis	(i)
-)	Right parenthesis	(r)
- *	Asterisk	(052)
- #	Pound sign	(043)
- $	Dollar sign	(044)
- ^	Caret	(136)
- ¦	Vertical bar	(174)
- \	Backslash	(134)
- ;	Semicolon	(073)
- <	Less than sign	(074)
- >	Greater than sign	(076)
- {	Left bracket	(173)
- }	Right bracket	(175)
- [	Open bracket	(133)
-]	Closed bracket	(135)

Table 12-2.   *NIST Encoding Mappings*

The Smidley Company

Ben Smith, President

1243 First Street
First Ciy, Texas 12345-1234
Phone: (111) 555-1212

X.400: G=Ben; S=Smith; OU1=HQ; P=Smidley; A=Telemail; C=us
Internet: Ben_Smith@smidley.com

Figure 12-2.   *X.400 address on a business card*

Ideally, you should be able to simply type this address exactly as shown and send a message to Ben Smith. However, some systems may use GN= instead of G= or may use a slash (/) instead of a semicolon (;) to separate the parameters. Adding to the complexity is that different ADMDs and PRMDs may require a different minimal set of parameters. For example, some may require an O= in the string whether it is needed or not.

Although many of these problems have been resolved for X.400 systems over the last few years, the industry has a long way to go in resolving all intersystem addressing problems. A single addressing convention to support all systems would enhance the use of electronic messaging across all companies and nations.

Transport

X.400 messages move through three distinct phases in an X.400 message handling system from the time it is sent to the time it is delivered. They are delivered and retrieved from message queues in the same manner as in SNADS, discussed in Chapter 9, "IBM SNADS and DIA." SMTP and Novell Global MHS use different methods to transport messages between MTAs. The following list defines the three X.400 transport phases:

1. *Message submission*   The MS or the MUA submits the message to the MTA using the P3 protocol.

2. *Message transport*   The message is transported from MTA to MTA using the P1 protocol.

3. *Message delivery*   The MTA delivers the message to the MS or MUA using the P3 protocol.

Message Submission

Before messages are submitted, the MUA, MS, or MTA (see Figure 12-1) must first log on to the appropriate component to establish its identity and the level of security to be used for the exchange. Logging on is called *binding*, and logging off is called *unbinding*. The initiating component may log on using one of two methods described in the standards: account and password, or token. The first method is well known to computer users. The second uses a public/private encryption key to identify each user. The second method is not widely used at this time, but will become more prevalent in the future.

An MTA-to-MTA logon and logoff uses BIND and UNBIND protocols, called MTA-BIND and MTA-UNBIND. Dialogues between the MTA, MUA, and MS use BIND and UNBIND to log on and off of components.

An important point about the BIND and UNBIND protocol is that only the component issuing the BIND can issue the UNBIND, as shown in Figure 12-3.

When a message has been built by the user (MUA), it is then submitted to the MTS either directly or through the MS. The message contains the P2 header and body,

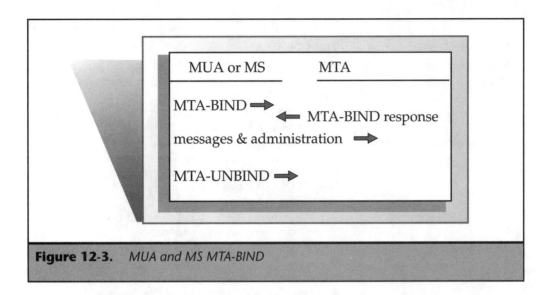

Figure 12-3. *MUA and MS MTA-BIND*

discussed later in this chapter. The only argument required in the P2 header is the IPM Identifier. The To:, From:, and other field types are optional. These missing fields occur in the P1 envelope the MTA constructs. When the message is passed to the MTA, a number of submission arguments are passed with the message that detail the kind of services the sender wants. These are services such as message delivery, blind carbon copies, and return delivery reports for all users.

Some of these message transport services are requested in the fields shown in Table 12-3, which are passed as the contents of the message submission envelope, a file that contains the normal e-mail fields seen in Novell Global MHS SMF-71 and SMTP/RFC 822 message formats discussed in Chapters 10 and 11. The X.400 envelope fields are listed in Table 12-3.

The P1 header used to transport the message is created by the MTA from the message submission arguments just described. Some locally stored arguments such as per-domain-bilateral-information, trace-information, and internal-trace-information are added to the header along with the responsibility fields added by the MTA. These additional fields are listed in the next section. The arguments are converted into the ASN.1 encoding scheme (discussed later in this chapter) before being placed in the P1 envelope. The P1 information deals with the recipients, the originator, and services such as priority, body part conversion information, delivery time, the type of reports requested, security issues, such as proof-of-submission or originators-certificate, and the type of contents.

Message Transfer

After the message has been submitted by the MS or the MUA to the MTA, the MTA will move the message toward the destination through other MTAs. The MTA first

Envelope Field	Field Contents
Message-originator	Name of person submitting the message
Message-recipients	Destination names and arguments that occur in fields such as To:, CC:, and BCC: message fields
Alternate-recipients-allowed	Argument that tells if the message can be auto-forwarded to other recipients specified by the sender
DL-expansion-prohibited	Arguments that tell if addresses that may be hidden in DLs can or can't be expanded. This is an important item. Some systems charge the originator on a per addressee basis. A DL that expands to many users adds unanticipated costs to the sender.
Disclosure-of-recipients	Argument that tells if the blind copy addressees are displayed to non-blind copy addresses
Priority	One of three values: normal, nonurgent, or urgent
Implicit-conversion prohibited	Parameter that tells if the contents of the P2 body can be converted to other forms. Conversion, such as from one word processing document to another, is one of the functions offered by some MTAs.
Conversion-with-loss-prohibited	If the conversion of the P2 body parts will result in the loss of data, such as highlighting or bolding, then the conversion will be prohibited
Deferred-Delivery-Time	The specified time in the future to deliver the message
Originator-report-request	Indication to the addressee's MTA the conditions under which a delivery report and a nondelivery report will be returned. The user may request that no reports be returned, only nondelivery reports be returned, or both types of reports be returned.
Content-return-request	Request for the MTA to return the content of the original message in the nondelivery report
Security-related	One of the following arguments: proof of delivery, proof of submission, or content integrity
Original-encoded-information-type	Argument that occurs in the header and indicates the type of information contained in the message body, such as Group 3 Fax or Undefined
Content-type	Definition of the structure of the message, such as 1984 IPM, 1988 IPM, Undefined, and EDI content types
Content	The message with the heading and body (the P2 IPM or P35 EDI message)

Table 12-3.   *X.400 Envelope Fields*

creates a session with the appropriate adjacent MTA by issuing the MTA-BIND and MTA-UNBIND commands. Once the session is created between two MTAs, the message is transferred from one to the other. After the destination accepts the message, the source MTA deletes it.

Arguments are passed between the two MTAs to tell the destination MTA how to handle the message. Most of these are passed to the MTA by the user and are derived from the submission envelope parameters listed previously in the section, "Message Submission"; others are only used between the MTAs.

Table 12-4 shows arguments that are only used between the MTAs. Other components such as MSs or the MUAs do not see these arguments.

The X.400 message is now enclosed in a series of envelopes containing appropriate information required by each component to move the message from the MUA through the MTA to the destination MTA, to be delivered to the end user's MUA. These nested envelopes are created and then removed by each component as the message is created, submitted, transported, and then delivered. The MTA creates a transport envelope composed of the arguments shown in Table 12-5. The MTA uses the contents of the Submission Envelope to derive its own envelope.

Message Delivery

When the final MTA receives the message for one of its registered MUAs or MSs, it initiates an MTS-BIND (shown in Figure 12-3) with the destination O/R-name entity to deliver the message or delivery report on the MUA or the MS. Arguments normally passed along with the P2 messages are the MTS-identifier, the delivery time, and the O/R-name or destination. Over thirty other delivery arguments may be passed in all. Some of these arguments are shown in Table 12-6.

MTA Transfer Arguments	Description
Message-identifier	This uniquely identifies the message across the entire system.
Per-domain-bilateral-information	This may contain one or many pieces of information.
Trace-information	This field documents the route the message took through all Management Domains on the way to the destination MTA.
Internal-trace-information	This field provides a history of the route the message took through the originating MTA.
DL-expansion-history	This field provides a view of all the distribution lists to which the message was sent.

Table 12-4. *X.400 Message Transfer Arguments*

MTA Envelope Arguments	Description
Originator-name	Identifies the account submitting the message
Recipient-name	Identifies the To:, CC:, and BCC: addresses
Originally-specified-recipient-number	Uniquely identifies the copy of the message on a destination address basis
Responsibility	Tells the destination MTA it has responsibility for the message
DL-expansion-prohibited	Prohibits hidden addressees in distribution lists from being used
Disclosure-of-recipients	Indicates whether blind copy addressees may be shown to nonblind copy addressees
Alternate-recipients-allowed	Allows messages to be routed to other recipients as specified by the sender
Priority	Contains one of three values: normal, nonurgent, or urgent
Implicit-conversion-prohibited	Tells whether the message body can be translated into other forms
Conversion-with-loss-prohibited	Prohibits translation with loss of data
Content-type	Describes message structure, such as 1984IPM, 1988IPM, undefined, or EDI
Content-identifier	Identifies the contents of the message; is generated by the MUA
Content-corelator	Identifies the relationship between the body parts of the message

Table 12-5. *MTA Envelope Arguments*

IPM Message Structure

The message is composed of two main parts, the envelope and the content. The Interpersonal Messaging (IPM) content, called *P2*, describes the structure, or what is inside the envelope. The content consists of the heading and the body. The body is further divided into one of several body parts. See Figure 12-4. Note that this envelope does not have a suffix as does the SNADS DIU envelope in Chapter 9, "IBM SNADS and DIA." Also, unlike the SMTP/RFC 822 and Novell SMF formats, which are in ASCII, the X.400 heading and body are in ASN.1 binary encoding, which is best left to computers to interpret.

Message Delivery Arguments	Description
Message-delivery-identifier	Uniquely identifies the message; derived from the MTS identifier and used to trace messages through the network; different from the IPM Identifier which belongs to the P2 envelope
Delivery-time	Indicates the time the message was delivered
Submission-time	Indicates the time the message was submitted
Originator-name	Indicates who the message is from
This-recipient-name	Indicates the name of the user receiving the message which may not be the original message addressee.
Intended-recipient-name	Indicates the name of the original addressee of the message, which may have been auto-forwarded by a message store
Redirection-reason	Tells why the message was redirected
Other-recipients	Indicates the members of the original To: and CC: fields
DL-expansion-history	Shows the DL expansion sequence, or DLs that expand to contain other DLs
Priority	Indicates the priority, which has one of three values: normal, nonurgent, and urgent
Proof-of-delivery-request	Requests that proof of delivery be returned
Content-type	Indicates is the type of message content, such as P2 or P35
Content	Indicates the actual message, including heading and body

Table 12-6. *Message Delivery Arguments*

ASN.1 Encoding

Of the four multiplatform messaging systems compared in this book, SMTP, X.400, SNADS, and NetWare Global MHS, two of them, SMTP and Global MHS, have message structures that are generally in plain English ASCII. The other two, X.400 and SNADS, are in encoded binary strings. Each binary string has three parts: type, length, and value. See Figure 12-5 for an example.

Abstract Syntax Notation is the notation (or encoded representation of the data) used to describe the X.400 and X.500 standards data types. It was first specified in the 1984 X.400 standard, X.409. It was later adopted by the Open Systems Interconnection (OSI) standard and expanded. The notation is composed of three general objects: values, types, and macros. Values always have a lowercase starting letter, types always have a capital starting letter, and macros are in all capitals.

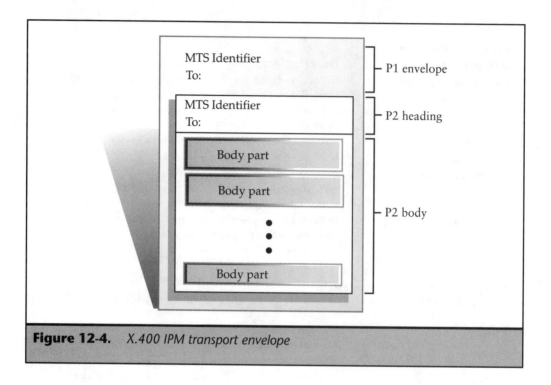

Figure 12-4. *X.400 IPM transport envelope*

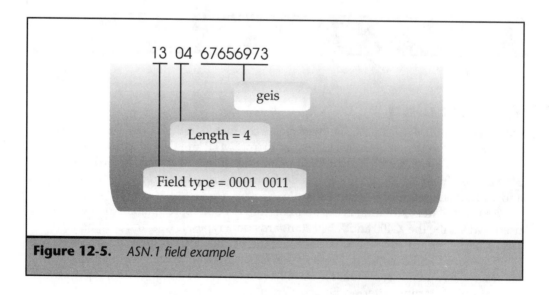

Figure 12-5. *ASN.1 field example*

There are many types defined in the ASN.1 encoding scheme. Some of them are shown in Table 12-7.

These types are translated into computer representations using the Basic Encoding Rules (BER) of the ASN.1 standard. These rules spell out in great detail how bits are transmitted, low-order or high-order first, how bytes (octets) are represented, and how each ASN.1 type is represented in bit stream.

The standard layout for representing a type is shown in Figure 12-6.

TYPES   There are four classes of types: universal, application-wide, context specific, or private. The first two bits on the left (bit 8 and 7) tell which of the four classes it is. The next bit (bit 6) tells whether the type is primitive or constructed of primitives. The last 5 bits and any necessary following bytes give the type's unique identifier number.

LENGTH   The Length field can specify the length of the construct or primitive, or contain a value, which means unspecified length. Those objects with unspecified lengths end in two bytes of all zero bits.

ASN.1 Types	Description
BOOLAN	True or false
INTEGER	Represents an integer
ENUMERATED	List of specific values
BIT STRING	A string of bits
OCTET STRING	A string of eight bit bytes called octets
NumericString	A string of numeric characters
PrintableString	A string of alphanumeric characters
IA5String	About equivalent to the US ASCII characters
UTCTime	In the format *yymmdd hhmmss* +/- 0000, where +/– 0000, is the offset from time zero, U.S.A. Central Standard Time is -0600
NULL	No value
OBJECT IDENTIFIER	The numeric object identifier
OBJECT DESCRIPTOR	The object title
SEQUENCE	Is a type that holds a sequence of other types
CHOICE	One of the type in the set
DEFAULT	No value is specified

Table 12-7.   *ASN.1 Encoding Types*

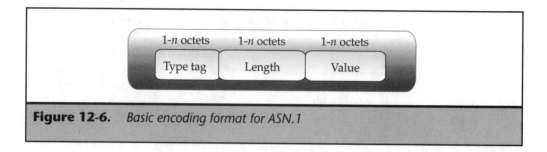

Figure 12-6. *Basic encoding format for ASN.1*

VALUE The value field can contain a value like the letter "a", a bit stream, a number or several types, either SEQUENCE, CHOICE, ENUMERATED or OCTET STRING. Thus a type may be a series of nested, recursive type specifiers. Logically, this is somewhat the same format as the IBM DIU packets in Chapter 9, "IBM SNADS and DIA."

An X.400 e-mail message could logically look like the ASN.1 encoded representation shown in Figure 12-7. The drawing is not a completely technical view but gives a valid logical representation of the data structure of the message. Each new rectangle is again represented in the type, length, value format shown at the top of the drawing. The fully capitalized words, such as SEQUENCE and SET, are values of the respective type fields.

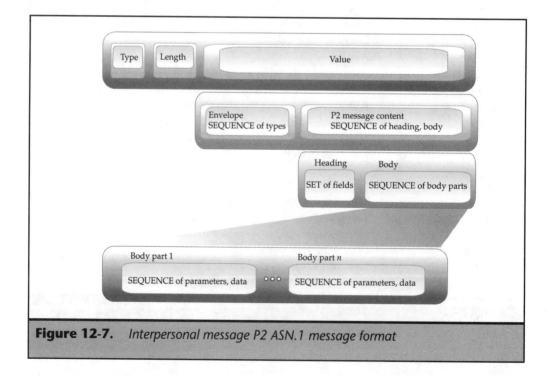

Figure 12-7. *Interpersonal message P2 ASN.1 message format*

Envelope

There are three types of envelopes in the P2 message: Submission Envelope, the Transfer Envelope, and the Delivery Envelope. The Submission Envelope is used to submit the P2 message to the MTA. The Transfer Envelope is used to contain the P2 message during MTA exchanges. The Delivery Envelope is used to deliver the P2 message to the MS or the MUA. Each of these is composed of somewhat the same information. A representative list of envelope fields was presented previously in this chapter.

Figure 12-8 shows an X.400 P2 envelope, heading, and body. Notice that it is a completely binary format and is not easily read without interpretation. Note the bold areas are the data fields. The data fields contain the values in order of presentation.

US
MARK400
MARK400.310000029402095539
us
mark400
quickcomm
peterson
tom
geis
02(P2)
US
MARK400 940209143513Z
us.mark400
quikcomm
autoanswer
geis

P2 Heading

The heading of an IPM message may contain several different information fields. These are encoded in ASN.1 and are not easily identified without being uncoded. The only one that must be in the heading is the IPM Identifier. The others are optional. Much of this information is submitted with the message to the MTA in the Submission Envelope. This duplication of information frequently is the reason that many of the P22 fields are not included in the message. One significant disadvantage is that because the MTA never touches the P22 heading or body (except to convert body parts as specified), the information in the heading is not touched. This means that since the BCC: field will appear both here and in the P1 envelope, the destination recipients will see who the BCCs are. Another disadvantage is that the IPM identifier is not used by the MTA for tracing a message. The MTAs do not look in the P22 envelope. Another unique message identifier, the MTA identifier, is created and used by the MTA to trace messages. The only required field is the IPM identifier.

The fields in the P2 heading are shown in Table 12-8.

P2 Body

The P2 body is a multipart body that may include text, binary, fax, telex, and so on in separate sections called *body parts*. Each section, as well as the whole message, is in ASN.1 encoding. ASN.1 encoding uses the first few bytes of the data stream to define the type, format, and length of the data. In this case, the start of each body part follows the last byte of the previous body part. The length of the previous body part is indicated in the length field of the ASN.1 string. For example, one of the ASN.1 objects in Figure 12-8 (shown in boldface type) is the ASN.1 encoding of the field interpretation shown in Figure 12-5. Each object has three parts to it: the type, the length of the value field, and the value, which may be of any length.

P2 Heading Fields	Description
IPM Identifier	Unique ID for the P22 message
Originator	Person sending the message (usually the same as the From: field)
Primary-recipients	To: field
Copy-recipients	CC: field
Blind-copy recipients	BCC: field
Subject	Topic of the message
Replied-to-IPM	ID of the IPM to which this message is replying
Authorizing-user	Person other than the originator who has the authority to issue the message
Related-IPMs	IPM Identifier of a related message
Obsoleted-IPM	ID of the message this one is making obsolete
Expire-time	Time and date after which the message is not good
Reply-time	Time by which the recipient should reply
Reply-recipients	Whom to reply to other than the originator
Importance	Importance rating
Sensitivity	Label given by the originator of the message, rating the message's sensitivity (personal, confidential, or private)
Auto-forward	Indicatation if the message was forwarded from some other user's MS

Table 12-8. *P2 Heading Fields*

```
a0 82 01 fd *[0] Length=509 (999995)
   31 81 ee *[Set] Length=238 (506)
         64 32 *[APPLICATION 4: MPDUIdentifier] Length=50 (236)
               63 11 *[APPLICATION 3: GlobalDomainId] Length=17 (48)
                     61 04 *[APPLICATION 1: CountryName] Length=4 (15)
                           13 02  [PrintableString] Length=2 (2)
                                 5553
                                 U S
                     62 09 *[APPLICATION 2: AdminDomainName] Length=9 (9)
                           13 07  [PrintableString] Length=7 (7)
                                 4d41524b343030
                                 M A R K 4 0 0
               16 1d  [IA5String] Length=29 (29)
                     4d41524b3430302e335c313030303030325c3934303230395c35353339
                     M A R K 4 0 0 . 3 1 0 0 0 0 0 2 9 4 0 2 0 9 5 5 3 9
         60 36 *[APPLICATION 0: ORName] Length=54 (184)
               30 34 *[Sequence] Length=52 (52)
                     61 04 *[APPLICATION 1: CountryName] Length=4 (50)
                           13 02  [PrintableString] Length=2 (2)
                                 7573
                                 u s
                     62 09 *[APPLICATION 2: AdminDomainName] Length=9 (44)
                           13 07  [PrintableString] Length=7 (7)
                                 6d61726b343030
                                 m a r k 4 0 0
                     83 08 [3] Length=8 (33)
                           7175696b636f6d6d
                           q u i k c o m m
                     a5 0f *[5] Length=15 (23)
                           80 08 [0] Length=8 (13)
                                 7065746572736f6e
                                 p e t e r s o n
                           81 03 [1] Length=3 (3)
                                 746f6d
                                 t o m
                     a6 06 *[6] Length=6 (6)
                           13 04 [PrintableString] Length=4 (4)
                                 67656973
                                 g e i s
         65 05 *[APPLICATION 5: EncodedInfoTypes] Length=5 (128)
               80 03 [0] Length=3 (3)
                     062000

                     —    —

         46 01   [APPLICATION 6: ContentType] Length=1 (121)
               02

                     —
         4a 0b   [APPLICATION 10: UAContentID] Length=11 (118)
               5541436f6e74656e744964
               U A C o n t e n t I d
         69 29 *[APPLICATION 9: TraceInformation] Length=41 (105)
               30 27 *[Sequence] Length=39 (39)
                     63 11 *[APPLICATION 3: GlobalDomainId] Length=17 (37)
                           61 04 *[APPLICATION 1: CountryName] Length=4 (15)
                                 13 02  [PrintableString] Length=2 (2)
```

Figure 12-8.   *IPM ASN.1 decoded message*

An ASN.1 IPM Encoded Message

A short IPM ASN.1 encoded message is shown in Figure 12-9. Note the boldfaced text: it is the data from which Figure 12-5 was derived.

Message Store Services

The message store (MS) is a new service provided to the message user agent (MUA) in the 1988 specification, and it functions like the post offices found on LAN e-mail systems. It interfaces the MTS on behalf of the MUA and stores messages for the user. Unlike the MUA, which is not always available to the MTA to accept messages, the MS is always available.

Some of the services the MS provides besides message storage are

Auto-forwarding messages
Deleting messages
Listing messages
Filtering messages

```
a08201fd3181ee643263116104130255536209130 7 4d41524b34
3030161d4d41524b3430302e335c313030303030325c39343032
30395c353533339603630346104130275736209130 76d61726b34
303083087175696b636f6d6da50f80087065746572736f6e8103
746f6da606130467656973650580030620
004601024a0b5541436f6e74656e74 4 4969469293027 6311610413
025553620913074d41524b34303031128 0 0d3934303230393134
333531335a820100a23e313c60333031610413027 5 7362091307
6d61726b34303083087175696b636f6d6da50c800a6175746f61
6e73776572a6061304676569 73 800101810200d0
04820108a08201043181e56b5060803080610413027 57 362091307
6d61726b34303083087175696b636f6d6da50f80087065746572
736f6e8103746f6da606130467656973000000001312514b2f30
312f39343032303931343353132a03c60803080610413027 573
620913076d61726b34303083087175696b636f6d6d
a50f80087065746572736f6e8103746f6da60613046765697300
000000a23d313ba0396080308061041302757362091307 6d6172
6b34303083087175696b636f6d6da50c800a6175746f616e7377
6572a6061304676569 7 30 0 0000000a8141412546573742 0 746f20
6175746f616e73776572 2 301aa01831038001051611
202020546f6d20506574 6 5 72736f6e0d0a
```

Figure 12-9. *IPM ASN.1 binary message*

One of the primary elements missing from the MS is foldering for the storing of messages. This is the ability to store messages in different folders and not as a long list of all messages, as is now the case.

The message user agent communicates with the MS by means of the P7 protocol. The MS, which sits between the MUA and the MTS, communicates with the MTA using the P3 protocol, the same protocol the MUA uses when it communicates to the MTA.

Each message in the MS is kept with information about the message. The MS has six ports with which to communicate with the MUA and the MTA for the receipt and transport of messages. The names of these ports are listed below. The first three are the ports which communicate with the MUA using the P7 protocol, the last three ports communicate with the MTA using the P3 protocol.

Message retrieval
Indirect message submission
Message store administration
Message delivery
Message submission
MTA administration

Address Directory

The 1984 X.400 standards did not specify a directory to contain X.400 addresses. It became evident that one was needed, but the individuals working on the 1988 X.400 standards believed the directory should support more than just e-mail related material. The ITU-TSS started another group to develop the OSI Directory to support multiple needs over and above those of e-mail. This set of recommendations was called the X.500 standard. Chapter 8, "X.500 Directory Systems," provides an overview of the X.500 standard.

In all likelihood, this standard will eventually support the Internet and X.400 messaging systems.

Summary

X.400 is an extensive standard whose 1988 and 1992 versions may handle complex binary documents and user service requests. The Interpersonal Messaging (IPM) portion of the service is one of two worldwide standards rapidly gaining acceptance. The other is the Internet SMTP. X.400 is more complex to define and to program than the SMTP and therefore has taken longer to reach the market. The 1988 standard is only now being implemented and on a limited scale, whereas new Internet standards are released and implemented three or four times faster. The X.400 standard is necessary for global exchange of electronic correspondence and other transactions.

Chapter Thirteen

OSI, LAN, and Other Protocol Stacks

As data communications networks proliferated worldwide, it became obvious that universally applicable standards for interconnecting those networks were needed. The International Organization for Standardization (ISO), a body of representatives from governments, vendors, trade and professional associations, and end users, generated the Open Systems Interconnection (OSI) standard, called X.200—a set of recommendations that provides a logical model for standardizing the development of new products and services used to interconnect computer networks. OSI pertains only to the interconnection of open systems, not the internal workings of computer systems.

Open systems, both hardware and software, are those that are nonproprietary and whose components are interoperable even though they may be from different and often times competing vendors. Open systems provide the consumer with more choices, better quality, and competitive pricing. In addition, applications developed on open systems are portable to other systems, thereby enabling customers to leverage their hardware and software investments.

The OSI architecture is a seven-layer protocol stack that enables the exchange of e-mail through the X.400 transport protocol. Each layer contains a group of similar functions required for the orderly exchange of data between participating systems. This layered approach shields the applications programmer from the details relating to the communications network. Another advantage is that changes made to one layer do not affect the other layers.

The X.400 Message Handling System (MHS) and the X.500 Directory System are application layer standards at the top of the OSI stack. Interconnection between different open systems requires an interface to the other system's six lower layers before e-mail can be successfully exchanged between the e-mail systems. This chapter reviews the role of each of the seven OSI layers in performing electronic message transfer. Figure 13-1 shows the four primary protocol stacks: IBM, Novell, OSI, and TCP/IP.

The OSI Protocol Stack

Before the ISO defined the OSI seven-layer protocol stack in 1976, interfacing systems entailed a complete programming project. Integrators had to start from scratch when connecting two different computer systems. Standard protocols for interconnection and standard routing codes did not exist. Code written to interface with one system, such as DECnet or SNA, often was unable to interface with another system. Also, recoding was often necessary when a protocol was upgraded. Network design was severely limited and inflexible.

The seven-layer OSI stack solves these problems by specifying what each layer of the data communications stack should do. All seven layers work together to transfer data between applications such as e-mail programs. Each layer (n) has knowledge only of the layer above it ($n + 1$) and the layer below it ($n - 1$). Also, each layer needs to communicate only with the layer above it and the layer below it—not all of the other layers.

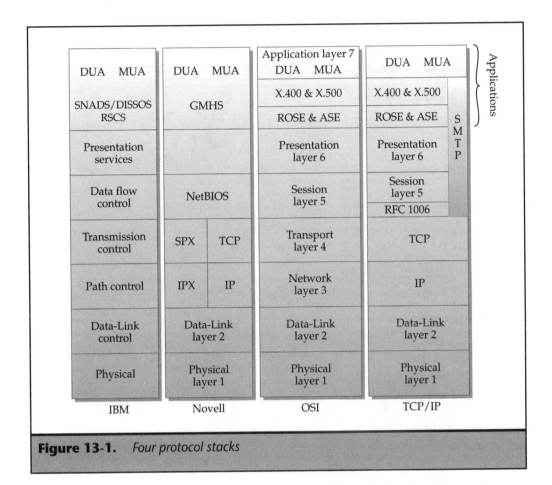

Figure 13-1. *Four protocol stacks*

Layer *n* of the stack provides services to layer *n* + 1 of the stack, and requests (and uses) services from layer *n* – 1 of the stack. A specific layer never communicates with other layers, such as *n* + 2 or *n* – 3. For example, the Network layer (layer 3), provides services to the Transport layer (layer 4), and requests services from the Data-Link layer (layer 2). The OSI model in Figure 13-1 shows the seven layers. A description of the function of each layer, from the highest layer to the lowest, follows.

Application Layer

The highest layer of the data communications stack is the Application layer (layer 7). This is the layer that requests services from the rest of the OSI stack, and is the layer on which all of the e-mail and directory application components run. The transport, message store, user agents, directory, and directory synchronizer all exist at this layer. The other layers in the stack exist to support the activities of this layer.

The Application layer receives input from the application itself and sends it down the stack to be transferred over the data communications network to the adjacent open system.

Presentation Layer

The Presentation layer (layer 6) offers services to the Application layer and requests services from the Session layer. It does not contribute functionally to most applications. This layer ensures that applications can successfully communicate even if they use different representations of the data.

Session Layer

The Session layer (layer 5) offers services to the Presentation layer and requests services from the Transport layer. It offers the following services:

- Orderly message exchange
- Checkpointing, synchronization, and status
- Conversation turns

It requests services of the Transport layer with calls such as Connect, Disconnect, Send Data, and Receive Data. This is the layer in which system logons and logoffs take place.

When you request services from another system and specify a system name, logon account, and password, you are requesting the services the Session level implements.

Transport Layer

The Transport layer (layer 4) is the lowest logical network layer. The Network layer (layer 3) is the highest layer that is involved with network hardware.

The Transport layer offers an end-to-end, error-free communications pipe for the layer above it, the Session layer. It also makes sure that the messages are sequenced properly. For example, suppose that a sizable file is transferred between two systems. The file is sent across the Network layer in three packets, each packet containing 512 bytes of data. After the packets are transmitted, the Transport layer ensures that they are reassembled correctly before they are presented to the Session layer.

An important option that occurs in standards at this layer, and significant to X.400, is the choice of one of five different levels of service. These service levels are called *transport classes (TPs)*, and range from the most basic and simple services offered (Class 0 or TP 0), to the most complex and robust (Class 4 or TP 4).

The only two that are ever really implemented are Class 0 and Class 4. Class 0 is used when layers 1, 2, and 3 offer error-free data to the upper layers. Protocols such as X.25, DECnet, and Systems Network Architecture (SNA) are examples of implementations that offer error-free data to the upper layers. Class 4 is used to ensure error-free transmissions when an implementation that does not offer error-free service, such as Ethernet, is used as the underlying data communications protocol transport. The TCP of TCP/IP fame is a layer 4, Class 4-type protocol and may be used without error-free services from the lower three layers.

The Transport layer requests services from the Network layer with calls such as Connect, Disconnect, Send Data, and Receive Data.

Network Layer

The Network layer (layer 3), controls the routing of data from the originating system through the network to the destination system. The Network layer and the remaining two, the Data-Link and Physical layers, complete the protocol stack. Layer 3 is the uppermost layer required for a system to participate in an OSI exchange. Systems with only the bottom three layers are called *intermediary* or *relay* systems. Data may have to travel through a variety of intermediate systems before reaching the final destination open system. The kinds of services that these three lower layers implement, such as X.25, Ethernet, Token Ring, and AppleTalk, are well-known names.

Data-Link Layer

The Data-Link (layer 2) implements protocols such as Ethernet IEEE 802.3 and Token Ring IEEE 802.5. It deals with how the media is used and how data frames are transmitted and received over the wire. Layer 2 provides error-free transmission across a physical link using the appropriate data-link protocol.

The frame structure is specific to the protocol type. However, the packets often have many of the same components, such as a frame start flag bit sequence, followed by source and destination addresses, control bits, a data portion, and an end flag bit sequence. The latter is often the same as the start bit sequence. The following diagram illustrates the structure of a data-link frame transported over an OSI network:

Start bit sequence	Source and destination addresses	Control data	The data	End bit sequence

The data portion carries the information from the Network layer. This data field also carries the headers attached by each layer as the frame has been transferred down the protocol stack.

Using the OSI stack, a package of data going across the network at the Physical layer would have the logical form shown in Figure 13-2. Many data transmissions have a header and a trailer as part of the envelope, which are not shown in Figure 13-2.

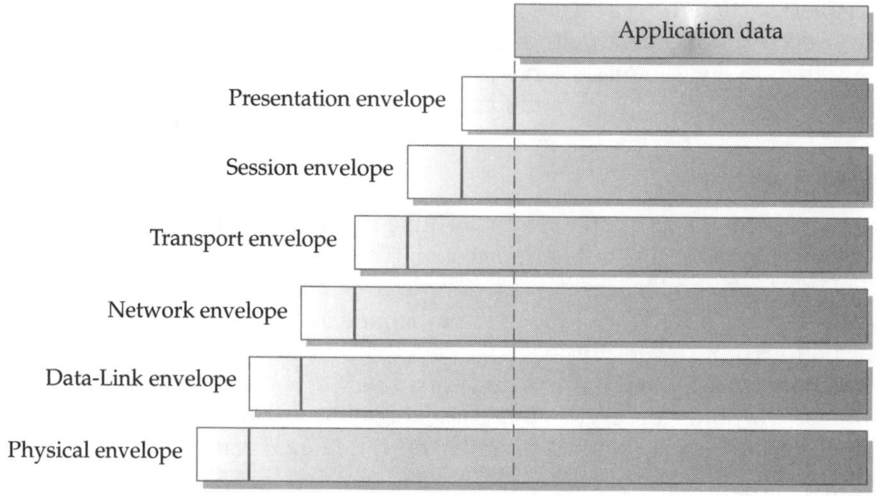

Figure 13-2. *OSI protocol stack transmission structure*

Each protocol type has different flags and a different packet structure, but all modern protocols follow the logic of this procedure.

Physical Layer

The Physical layer (layer 1) provides for the transparent transmission of bit streams across a physical connection. The Physical layer implements X.25, ISDN, and LAN protocols. These types of protocols deal with the electrical part of the connection, with the modem, and with the electrical interface, such as RS-232. The Physical layer is responsible for establishing the logical connection over the physical media, such as twisted-pair, fiber, and coaxial cable, and wireless media. This logical link can be established and released without affecting the physical line connection.

Summary

In most e-mail integration projects, a detailed list of what operating systems and protocols are supported is supplied with the product. These usually name the protocol stack and the bottom three layers, such as Token Ring or X.25. X.400 was originally defined as a layer 7 application on the X.25 stack only. It has just recently been able to support Ethernet, TCP/IP, and Asynchronous (PC dial-up) transmission.

Chapter Fourteen

E-mail Network Management

If data communications network management is a real concern, e-mail network management is even more problematic. For years, companies have attempted to put in place the management architecture and services necessary to maintain an efficient and standard level of service for their data communications networks. Yet it is increasingly obvious, especially to Internet users, that e-mail networks have become larger than the underlying physical networks. Since e-mail is by nature a store-and-forward process, it assumes that not all network addresses of e-mail users are continually available on the data communications network. Also, because of dial-up protocols like UNIX-to-UNIX Copy Program (UUCP), and Serial Line Internet Protocol (SLIP) on the Internet, the e-mail world spans the physical device addresses of all the data communications networks and personal computers with modems. Though e-mail applications cannot exist without underlying layers and management of their networks, an e-mail application is a network in its own right—a more user-visible network than the underlying data communications network. E-mail management tools are rapidly becoming a necessity for operating any e-mail network.

E-mail Management Standards

In the recent past, organizations were primarily concerned with connecting all of their disparate e-mail systems. Just finding creative ways to get all of the systems to exchange messages reliably with one another was enough of an undertaking. Today, however, through the use of gateways, most organizations can consolidate all of their e-mail systems into an interoperable network. With this high level of connectivity come additional problems, though: end-to-end management of large e-mail networks, monitoring of traffic flow, configuration management, disaster recovery, and a myriad of other related issues that keep e-mail network administrators up at night.

Messaging industry groups such as the Electronic Messaging Association (EMA) launched subcommittees to gather input for the development of e-mail system and network management tools for their member companies. Also, a consortium of Isocor, Microsoft, and other leading messaging industry vendors recently launched the Messaging Management Council (MMC) to investigate e-mail management problems and apply existing and planned technology to develop e-mail network management tools. Their scope will be backbones, servers, gateways, clients, converters, and public networks.

Other organizations are in the process of developing standards for the management of open systems and networks. The International Organization for Standardization (ISO), the International Telecommunications Union (ITU), the International Federation of Information Processing (IFIP), and the Internet Engineering Task Force (IETF) are the primary organizations involved in the creation of e-mail network management standards.

International Organization for Standardization (ISO)

The ISO created the Open Systems Interconnection (OSI) seven-layer stack so often seen in the data communications world. It released a specification called OSI Management Framework (ISO 4798-4), which details five areas that must be covered in any network management facility. These are also necessary in the e-mail network, and each is covered later in this chapter. In the area of e-mail network management, the ISO and the telecommunications standardization sector of the ITU are all working together under joint efforts to define and develop management models.

International Federation of Information Processing (IFIP)

The second major effort for the development of e-mail management tools is being conducted by the IFIP, which first released an e-mail requirements document in January, 1992. That document is one of the more active efforts in the area of e-mail management, and its latest release, dated October, 1993, cites such e-mail management topics as core management functions, naming, addressing, routing, message logging and tracking, disaster recovery, rate control, and management information exchange. In addition, the document addresses the same topics—configuration, fault, performance, security, and accounting management—that appear in the ISO Data Communications Management Framework specification (discussed in this chapter in the section, "E-mail Management Framework").

Internet Engineering Task Force (IETF)

The IETF is the primary standards-making body for the Internet. Standards are reviewed and released through a process called Request For Comment (RFC). Much of the work is done online through e-mail correspondence using distribution lists. The IETF has been very active in the areas of the Simple Network Management Protocol (SNMP), Management Information Base (MIB), and e-mail management. The IETF's management workgroup is called Mail and Directory Management (MADMAN).

E-mail Management Framework

The ISO established the Data Communications Management Framework specification in ISO 4798-4 that covers five areas to address in any open system implementing the OSI seven-layer protocol stack. This model defines the general categories needed to implement e-mail management. So far, ISO's e-mail management effort has focused on items called objects. *Objects* are stand-alone entities—information units that can be named—like the ASN.1 primitives or constructs (such as COUNTRY or SURNAME), discussed in Chapter 12, "X.400 Interpersonal Messaging." An object generally has three components: type tag description, length, and value. The five areas of network or system management described in the ISO documents are accounting management, fault management, performance management, configuration management, and security.

ACCOUNTING MANAGEMENT   Management of resource use, which includes collecting and processing data on the number and size of e-mail messages, number of addresses, and other resource use. It also allows the limiting of resources such as message size, disk space, and number of files. Statistics and accounting data can be generated at the user and resource level. Users are assigned quotas for individual resource usage.

FAULT MANAGEMENT   Involves reporting, tracking resolution, detection, isolation, and repair problems. This is the trouble reporting system often used by large network or system management organizations. It includes problem identification tools to isolate, diagnose, resolve, track, and report problems.

PERFORMANCE MANAGEMENT   Collecting and analyzing data to evaluate the performance of the system. This could include modeling as well as other data analysis tools. This information can be used in capacity and traffic flow planning efforts.

CONFIGURATION MANAGEMENT   The initiation, management, and closing down of management components. It can also be the collection of data in order to determine when changes are made in the system.

SECURITY   Services that provide component authentication, key distribution, access control lists, choices for how users log on, and what resources they have a right to use.

Management Information Base (MIB)

Most e-mail management efforts focus on *management information bases (MIBs)*. A MIB is a set of objects contained within an open system repository. It stores information used to control the management of an open system with the use of protocols. MIBs are used to manage such system components as remote operations and configuration management. The ISO, the IFIP, and the Internet are active in defining MIBs. The Internet is the most active in the e-mail area. The IETF just released an e-mail MIB in RFC 1566. Specifications for MIBs for other network components, such as the Domain Name Service (DNS), the X.500 directory system agent (DSA), and Ethernet are also being drafted.

The MIB defined in RFC 1566 is specifically designed to be associated with the e-mail message transfer agent (MTA). The RFC 1566 uses a modified form of ASN.1 to describe the data objects of the MIB. Some of the information fields in the MIB are number of characters, number of addresses, and number of messages transferred through an MTA. Once these information fields are defined and implemented by the platform on which the MTA exists, any management system can read the structure and gather information stored by the system in the MIB. This information can be used by the performance management, accounting management, fault management, and configuration management subsystems, discussed earlier, to manage the e-mail system.

The following table lists some of the items in the MTA MIB defined by RFC 1566.

MIB Item	Description
Received messages	Number of messages received since the last MTA initiation.
Stored messages	Number of messages currently stored in the MTA.
Transmitted messages	Number of messages transmitted since the MTA was initiated.
Received volume	Received volume of data in thousands of bytes. This should only include the message heading and body data, not the envelope data.
Stored volume	Number of thousands of bytes currently stored in the MTA.
Transmitted volume	Transmitted volume of data in thousands of bytes. This should only include the message heading and body data, not the envelope data.
Received recipients	Total number of addressees moving into the MTA.
Stored recipients	Total number of addressees currently stored on the MTA.
Transmitted recipients	Total number of addressees moving out of the MTA.

Each MIB can be read a row at a time to retrieve information. Figure 14-1 shows the logical structure of a MIB. Each row represents an object type and its three components: length, type description, and value.

Field Type	Length	Value
Total characters		
Total addresses		
Total messages		
⋮	⋮	⋮

Figure 14-1. *Representation of an e-mail MTA MIB*

Summary

E-mail will be widely used for global business correspondence only when network management technologies stabilize present implementations. Currently, it is very difficult to trace e-mail problems across multiple networks or analyze performance bottlenecks. E-mail messages going to the same destination may take 30 minutes or two days to be delivered. These problems will gradually subside and delivery times will become shorter and tighter as e-mail management tools help find, analyze, and solve the aberrations that occur on a global e-mail network.

Chapter Fifteen

E-mail Gateway Options

E-mail is used in many organizations. It is more important than the ubiquitous telephone, because it can convey the primary information necessary for many business processes, which are largely what define an organization.

Corporations utilize a large number of unbudgeted resources to purchase and manage their current disparate e-mail systems. They recognize the strategic importance of e-mail in today's business environment and want to integrate the multiple systems into a reliable service. Depending on the number and types of e-mail systems and the number of users, this integration can cost millions of dollars. This chapter presents the most feasible options for implementing gateways to integrate different e-mail systems, and provides a comprehensive comparison of gateway products.

Overview

A plethora of e-mail products for integrating a company's existing e-mail systems into a homogeneous enterprise-wide network are available on the market. Existing e-mail systems range from historic mainframe systems, such as IBM's PROFS and Verimation's MEMO, to mini-based systems like DEC's ALL-IN-1 and HP's DeskManager. They include SMTP systems, Novell based e-mail packages like Futurus' Right Hand Man, and major LAN packages like Lotus' cc:Mail and Notes, Microsoft's Microsoft Mail, and WordPerfect's WordPerfect Office.

To deal with the complexity of e-mail integration, The Drummond Group, of Fort Worth, Texas released an e-mail integration Request For Information (RFI) to eighteen primary vendors in the industry. Sixteen vendors responded. The RFI focused on e-mail integration products rather than integration services and addressed three broad categories: mail transport, gateways, and directories.

The vendors included in the RFI were well-known companies such as DEC, Novell, NCR, and HP, and e-mail specialists such as Worldtalk, Nexor, and ISOCOR. Vendors that offer an integration service were asked to supply information only on the primary e-mail systems for which they offer integration products. There are currently over 50 e-mail system products on the market. The RFI asked for information on the most commonly implemented products, such as ALL-IN-1, Banyan VINES, HP DeskManager, Novell NetWare Global MHS, Lotus Notes, Lotus cc:Mail, Microsoft Mail for the PC, Microsoft Mail for the Macintosh, IBM OfficeVision products such as OV/VM, OV/MVS, OV/400, SMTP, WordPerfect Office, X.400 1984 and 1988 standards, and X.500.

The products generally fall within several categories. In many cases the vendors offer products that fit or span a number of these categories:

Stand-alone gateways
Multigateway hubs
Multigateway distributed switches

Directory and transport backbone components

Product-independent message store products

Stand-alone gateways often work best for connecting one, two, or three e-mail systems, while multigateway hubs work best for connecting e-mail systems in more complex and geographically dispersed environments. Multigateway distributed switches and backbones work best for geographically distributed, multi-e-mail system integration projects. The distinction between different systems often becomes blurred as products and marketing strategies evolve.

Stand-alone Gateway Solutions

Retix and Worldtalk historically have offered a variety of stand-alone gateway solutions, though both are rapidly moving toward multigateway switches. Other firms offering stand-alone gateways are Microsoft, StarNine, and Computer Mail Services, Inc. Keep in mind when mixing and matching gateways from several different vendors, it is important to recognize that each of them:

■ Will use different management tools

■ Will use different directory solutions

■ May not integrate well

■ May require unanticipated people resources to manage

Figure 15-1 depicts four different e-mail systems interconnected in a point-to-point arrangement by six stand-alone gateways. Addition of even one more e-mail system will result in growth to ten gateways.

These stand-alone gateway solutions frequently are implemented on a single Intel processor using DOS, UNIX, or OS/2 as the operating system.

Multigateway Switch Solutions

AlisaMail, Wingra's Missive, and Soft*Switch Central offer multigateway switch solutions like the one shown in Figure 15-2. These products have a single hardware hub with several well-integrated gateway modules. They usually have a centralized address conversion utility, a directory synchronization utility, solid management tools across all components, document conversion facilities, and various plug-and-play e-mail protocol conversion module options. Unlike distributed multigateway switches, in which the gateway modules are interconnected, distributed components of the switch, nondistributed multigateway switches contain gateway modules that connect to a single machine at a single location.

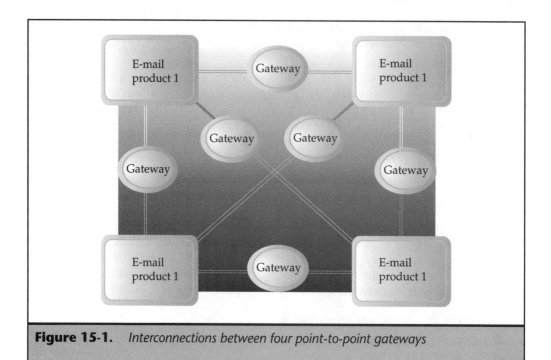

Figure 15-1.   *Interconnections between four point-to-point gateways*

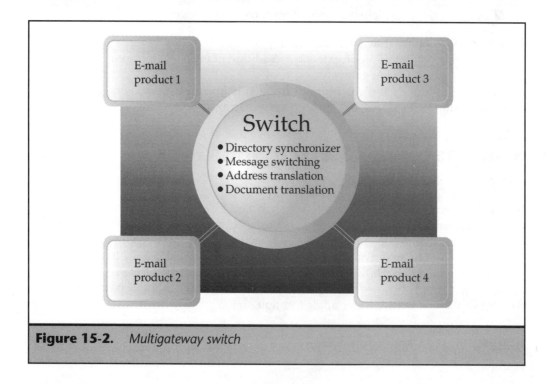

Figure 15-2.   *Multigateway switch*

Multigateway Distributed Switch Solutions

Examples of multigateway distributed switches are Soft*Switch's Enterprise Messaging Exchange (EMX), The Boston Software Works' InterOFFICE Message Exchange, CompLink's (formerly Technology Development Systems) NetSwitch, Unified Communications' Unified Networking Systems, DEC's Enterprise Messaging Services, NCR's StarPRO Enterprise Messaging, HP's OpenMail products, and ISOCOR's ISOPLEX, ISODIR, and ISOGATE products.

This category is a distributed version of the previous gateway type, implementing the same functionality over multiple, coordinated, distributed hardware components, as shown in Figure 15-3. These separate components intercommunicate by means of a directory and transport system. Frequently, these are based on SQL and X.500 platforms for directory exchange and on X.400 and SMTP for interswitch message exchange.

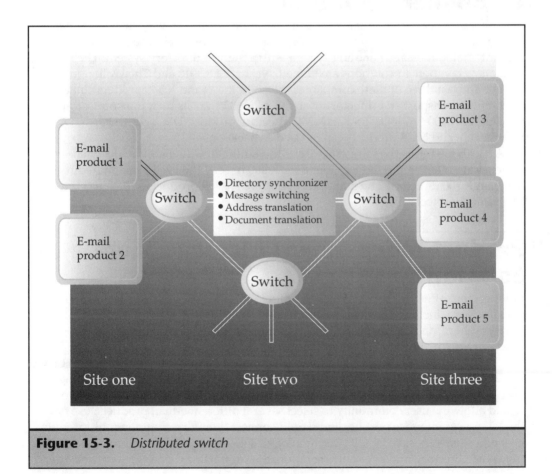

Figure 15-3. *Distributed switch*

Worldtalk's marketing strategy focuses on using existing X.400 and X.500 platforms from other vendors. Thus Worldtalk implements the directory synchronization and gateway functionality, but depends on other vendors for the transport and central directory.

All of the vendors in this category have marketing or strategic alliances with many other vendors in the first three categories. They often depend on companies such as Worldtalk, The Boston Software Works, ISOCOR, and Retix, for part of their total product solution. This is especially true in the area of e-mail gateway technology. In the product matrix, shown at the end of this chapter in Table 15-1, many of the indicated third-party gateways are actually integration solutions from one of these four vendors.

Directory and Transport Backbone Component Solutions

The difference between this category and distributed multigateway switches is that the vendors in this category offer a *backbone* infrastructure that is a robust and well-integrated message handling system with a built-in directory system. The major vendors of component solutions are Novell, Nexor, ISOCOR, and The Wollongong Group. The last three of these companies sell solid X.400 and X.500 products that form very robust transport and directory systems. Novell offers its own transport protocol, third-party X.400 products, and an X.500 directory product. DEC, HP, CDC, Soft*Switch, and NCR products can be used as backbones.

A backbone forms the infrastructure of the multigateway distributed switches. The switches are often connected together over X.400 or SMTP, with the common directory being an X.500 directory. The main exceptions are Novell products, which use the MHS transport technology.

Product-Independent Message Store Solutions

Products that offer independent message store functionality, not tied to one specific e-mail system, include Hewlett Packard's and DEC's products for cc:Mail and Microsoft Mail, and NCR's NCR for Microsoft Mail. ISOCOR also offers this functionality for 1988 X.400 Message Handling Systems.

Corporations are attempting to establish a uniform messaging infrastructure that will allow them to reduce management and operation costs, while still supporting the user's needs with respect to product selection. Independent message store technology would allow the user community to select several different e-mail products, use only the MUA user interface portion, and would connect them to a common multiproduct post office (*message store* in X.400 terminology). The result would be a uniform set of transport, directory, and post office components supporting a diverse set of end-user selected interfaces such as cc:Mail, Microsoft Mail, and others. This configuration offers the possibility of reduced network management costs, while enhancing service and availability. Currently, these architectures only support cc:Mail and Microsoft Mail, but there will certainly be more e-mail platform products to come in the future.

Wingra's Missive and Alisa operate on existing DEC platforms. Boston Software Works' InterOFFICE Message Exchange has software actually implemented on each platform, foregoing the need for additional PC gateway hardware. Novell's offering, the MHS, uses mostly third-party gateway technology as a backbone transport. Retix has the largest share of the e-mail gateway market and sells products both to end users and to other vendors for implementation in their products. Unified Communications' UNS distributed switch supports X.400, SNADS, Global MHS, and SMTP.

Interpreting the Product Matrix

There are several functionality areas you should be aware of when selecting a product for an e-mail environment. Table 15-1, located at the end of this chapter, compares products from several vendors. The following are descriptions of the categories represented in the table and its legend.

NON-X.400 GATEWAY WITH BINARY SUPPORT   Support for both textual and binary attachments through the gateways have become a standard feature of e-mail system exchange over the last year. In Table 15-1, the letter "x" indicates support for both textual and binary attachments.

SUPPORT FOR X.400 1984, 1988, AND 1992 VERSIONS   The feature set of the X.400 1988 and 1992 versions is much broader than the 1984 standard. The biggest difference, from a gateway point of view, is that the 1988 and 1992 versions support several documents in a single message. They downgrade to support services of earlier versions, so companies need only implement a higher version in order to support lower level versions. In Table 15-1, "84," "88," and "92" indicate vendor support of the X.400 standards for the 1984, 1988, or 1992 releases, respectively.

THIRD-PARTY GATEWAY WITH BINARY SUPPORT   Some degree of integration with third-party products usually exists as a result of the numerous X.400 and SMTP gateways available on the market. However, some of the products work better together than others, and have better integration and manageability across all components of the platforms. The word "certified" is used loosely to mean that the primary vendor has a solid working relationship with the third-party vendor and that their products work well together. In Table 15-1, the number "3" denotes third-party solutions.

TUNNELING BETWEEN LIKE SYSTEMS   With the advent of the e-mail switch to interconnect regional and campus e-mail environments, the same products implemented at different sites don't necessarily communicate directly with just their own protocols. They communicate through the switch architecture used to integrate them with other e-mail systems. Switches naturally translate between e-mail systems, often converting inbound message data to some neutral format before reconverting it for the outbound system. Tunneling functionality bypasses interswitch translations and modification, tunneling the message to the other end without alteration. In Table 15-1, the letter "t" indicates that the switch, hub, or backbone offers tunneling between similar systems.

DIRECTORY SYNCHRONIZATION AND ADDRESS TRANSLATION   Directory synchronization is an important feature of transparent e-mail integration, because it is the basis for implementing transparent user addresses between systems. For example, a user on HP DeskManager addresses a message to a user on WordPerfect Office using the addressing format on the HP DeskManager system. Directory synchronization then translates and updates all proprietary directories so that addresses of users on other systems appear as native. In Table 15-1, the letter "d" indicates that the vendor's directory synchronization product supports that particular e-mail system.

SUPPORT FOR MIME OVER SMTP   Internet mail is a text-based e-mail system that, with the addition of the MIME standard, now supports nontext-based multiple bodyparts. MIME makes the SMTP network similar to the X.400 1988 functionality, with respect to multiple bodyparts being supported within a single message. In Table 15-1, the letter "m" indicates that the product's SMTP interface supports MIME.

X.500 SERVICES   All of the switch vendors implement a central X.500 directory or an SQL database, except Worldtalk, which deliberately uses other vendors' architecture. The central directory is necessary to implement address translation, document conversion, and sophisticated message routing. Central directories also form the basis for companywide and worldwide information retrieval analogous to white and yellow pages. Many corporations are implementing online phone directories and demographic databases for their employees based on X.500.

The 1993 X.500 standard has just been published, but many vendors will not have the 1993 functionality incorporated in their products until the end of 1994 or later. Some vendors released 1988 versions with "1993-like" extensions, such as access control to records and fields within the X.500 directory and shadowing between the different parts of the distributed directory. *Shadowing* is used to update other parts of the distributed X.500 database with information from the primary information owner. In the X.500 column in Table 15-1, "88" means a product has 1988 functionality and "93" means a product has 1993 functionality.

SUPPORT FOR DOCUMENT CONVERSION   If a company supports document conversion, it almost always uses KEYpak from Keyword Office Technologies Ltd., of Canada. A couple of products support MasterSoft, Aladdin, and other proprietary conversion utilities in addition to KEYpak.

SUPPORT FOR RFC 1006 FUNCTIONALITY   The RFC 1006 Internet standard defines how OSI protocols can be carried over the TCP/IP Internet protocol stack. RFC 1006-compliant systems allow OSI layers 5, 6, and 7 to use TCP/IP layers 3 and 4. Since most corporations are using TCP/IP and OSI, this feature is necessary to allow the underlying protocol stack independence from X.400 and X.500 to facilitate its widespread use.

Interpreting Blank Intersections

A blank intersection in Table 15-1 means that the e-mail system does not offer its own internally developed gateway or does not offer a gateway from a certified third party. However, such systems are capable of communicating with the product through the product's X.400 implementation.

Summary

Many fine products exist for integrating several e-mail products within an organization, but it is an expensive project. It is therefore essential that you select the products that best suit your needs. By using the tables and figures provided in this chapter, you have the opportunity to evaluate both your requirements and what is available in the marketplace.

Company	Product	DEC ALL-IN-1	Banyan VINES	HP Desk-Manager	IBM SNADS	Novell MHS	Lotus Notes	Lotus cc:Mail
Alisa Systems, Inc.	AlisaMail	x,d				x,d		x,d
The Boston Software Works, Inc.	InterOFFICE Message Exchange	x,t,d	x,t,d	x,t,d		x,t,d		x,t,d
Control Data	Mail*Hub	x,d			x,d	x,d	x,d	x,d
Digital Equipment Corporation	MAILbus	x,t,d	d	88,t,d	x,d	3	3	x,t,d
Hewlett Packard	OpenMail, X.400, X.500, SMTP	88,t		x,t	3	3	3	x,t,d
ISOCOR	ISOGATE 400, ISOPLEX Message Server, ISOPLEX DX, X.400 Lite		3	3	3	x,d	x,t,d	3
NCR	StarPRO Enterprise & Microsoft Mail for StarGROUP	88,t,d	x,t,d	x,t,d		x,t,d	x,t,d	x,t,d
Nexor	XT-PP, XT-QUIPU, XT-DUA, XT-MUA, PC-DUA						3	3
Novell, Inc.	NetWare Global MHS with Products and Gateways	3	3	3	3	x,t,d	3	3
RETIX	OPENserver 400, MH-5420, MH-6000 and MG-xxx Gateways		3		3	t	t	t
Soft*Switch	Enterprise Mail Exchange and Central	x,t,d	x,d	x,d	x,d	x,t,d	x,t,d	x,t,d
CompLink	NetSwitch	x,t	x,t	x	x,t,d	x,t	x,t,d	x,t,ds
Unified Communications Inc.	Unified Networking Systems (UNS)	d,88			x,d	x,d		x,d
Wingra Technologies, Inc.	Missive	x,d				x,d	x	x,d
The Wollongong Group, Inc.	PathWay Messaging Services							
Worldtalk	Worldtalk 400				3	x	x,t	x

LEGEND

Non-X.400 Gateway with binary support	x
X.400 (84)	84
X.400 (88)	88
X.400 (92)	92
Third-party Gateway with binary support	3
Tunneling between like systems	t
Directory synchronization	d
MIME supported	m

Table 15-1. *Product Comparison Matrix*

Microsoft Mail for the PC	Microsoft Mail for the Mac	IBM OV/VM	IBM OV/VMS	IBM OV/400	SMTP	WordPerfect Office	X.400	X.500	Document Conversion	RFC 1006 OSI over TCP/IP
x,d	x,d				x,d	x,d	84	None currently	Yes	Yes
x,t,d		x,t,d		x,t,d	x,t,d		84,88	None currently	Yes	Yes
x,d	x,d	x,d	x,d	x,d	x,d	x,d	84,88	1993 Almost	Yes	Yes
x,t,d	x,t,d	x,t,d	x,t,d	d	x,t,d,m	3	84,88	1993 Almost	Yes	Yes
x,t,d	3	3	3	3	x,t,d	3	84,88	1993 Almost	Yes	Yes
3	3	3	3	3		x,t,88	84,88,92	1993 Almost	No	Yes
x,t,d	x,t,d	3			x,t,d,m	x,t,d	84,88	1993 Almost	No	Yes
					x,m		84,88,92	1993 Almost	Yes	Yes
3	3	3	3	3	x	3	84,88	None currently	Yes	Yes
t,d	t				3	3	84,88	1988	No	No
x,t,d	x	x,d	x,d	x,d		x,t	84,88	None currently	Yes	Yes
x,t,d		x,t,d	x,t,d	x,t,d	x,t	x,t,d	84,88	None currently	No	Yes
x,d	3		x,d	x,d	x,d,m	x,d	88	1988	No	Yes
x		x	x	x	x,m	x,88,d	3,84,88	1988	Yes	No
					x		84,88	1988	No	Yes
x,t,d	x,d	3	3	3	x	x	No	None	No	Yes

Chapter Sixteen

Document Conversion Options

When you e-mail a locally generated computer file that includes an attachment such as a word processing document, a spreadsheet, or a graphic file to users on a different e-mail system, you want those files to be readable in the recipient's own application programs. Vendors such as Keyword, MasterSoft, Aladdin, and DataVis seek to resolve the formatting problems of document conversion. The most popular document converter in the e-mail environment is KEYPak, which is used in this chapter to describe an example of state-of-the-art document conversion technology.

A recent development in document conversion software is the use of document readers. A *document reader* differs from a *document converter* in that a reader creates a noneditable, yet readable and printable file, while the converter creates an editable, revisable file. The reader is invoked by the user on the MUA, whereas the converter can translate the document transparently without the end user's knowledge of the conversion. Document readers are likely to be more popular than converters in the future, because they are easier to implement and because the end user is responsible for executing them, rather than the network.

Document Conversion and E-mail

Document conversion facilities in the e-mail environment are provided on either the MUA, the MS, or the MTA. On the MUA, the user converts outgoing and incoming documents to the appropriate format. This gets very cumbersome if the outgoing document has more than two addresses on the distribution list. A separate message with the appropriate destination format has to be created for each addressee's application. In the MS and MTA, conversion facilities are automatically invoked.

Most document conversions are performed within the MTA, unless the user restricts the conversion during message submission. Before converting from the source format to the destination format, the MTA must have the following information in order to perform the conversion:

■ **The source document format**

Most MTAs—at least the X.400, SMTP, SNADS, and Novell Global MHS MTAs—include a field in the message structure that identifies the format of a document. If the document type is identified, then one piece of information necessary for automatic document conversion is in place; the recipient of the document will be able to identify the particular document format.

■ **The destination document format**

The MTA needs to have information about the format of the destination user's application. This information generally is entered when the user registers with the e-mail system's MTA or directory system agent (DSA). As X.500 directories become more prevalent, information on the user's preferred format will be included in the user's directory entry as one of the attribute values associated with that user's unique X.400 address. See Chapter 8, "X.500 Directory Systems" for more information.

- **Sender preference for document conversion if no information will be lost**

    Each sender may specify in the message structure whether the document should be converted if no information, such as text highlighting or boldfacing, will be lost. Also, details of the kind of document information that will be lost during conversion must be known so that decisions can be made regarding conversion. For example, it may be appropriate to convert a document if something minor, like a specific font type, will be lost, but not if something considered more important, such as boldfacing, will be lost.

- **User permission for document conversion**

    Users must tell the initial MTA whether they want conversion. This information is carried in the message envelope so that all MTAs will convey the user's wishes.

    In some e-mail systems, all MTAs are able to convert documents as a service to the user. In others, only a few MTAs support document conversion. Other MTAs route messages with document conversion needs to MTAs with document conversion capabilities. How document conversion is implemented is specific to the requirements of the particular message handling system. Issues such as implementation costs, performance overhead, and anticipated document conversion frequency often determine where and how a company will institute document conversion capability.

Example: E-mail Document Conversion Process

User A creates a message in Microsoft Word for Windows format, indicates to the MUA it is a Word for Windows document, specifies that the document should not be converted if information will be lost, and attaches it to the message, addressed to User B. User B uses the WPS-Plus application from DEC. The MTA builds an envelope around the e-mail message containing the Word for Windows document, notes the document type in the envelope, and indicates that conversion should not take place if it will cause information to be lost. The MTA passes the envelope on to the next appropriate MTA. Between sending and receiving the message, one of the MTAs registers that the sender and recipient have different word processor applications and that the message can be converted if no information loss will occur. Then that MTA either converts or doesn't convert the message to the destination type, depending on whether its table shows that information loss will occur between the two document types.

> **What Constitutes Information Loss?** Everything in the previous
> example seems straightforward until you recognize that in some instances
> Word for Windows may be completely convertible to WPS-Plus and in
> others, it may not. For example, suppose the Word for Windows document
> has a table built into the text describing revenues by month and WPS-Plus
> does not support the table command. If the document converter can create
> something that approximates the original table in the WPS-Plus document, is
> that loss of information or not? Should the document be converted? Except
> for the table, the original document could be converted with complete
> fidelity.
>
> The MTA does not have the information necessary to decide whether to
> convert the document or not. Does this mean that the MTA must pass every
> document that needs conversion to the converters to determine whether loss
> of information will take place? To determine this, document conversion
> without loss should be defined for each document format. When choosing to
> implement document conversion capability, look closely at the actual formats
> the software will convert, and investigate its limitations before selecting and
> implementing it.

Document Conversion

The KEYPak software program from Keyword Office Technologies converts word
processing documents that include picture formats. The fidelity of the original is
maintained through the use of a neutral document format called ODX, which enables
the exchange of documents without loss of information. Sometimes information that is
not content specific, such as tables or character sets, may not be available in the
destination format.

Features not supported in the destination document are treated in one of three ways:

- They are converted to a structure that mimics the source document feature.
- No conversion is made.
- A warning is given indicating that a nonrepresentable structure in the
  destination system was encountered.

The types of document conversion supported by KEYPak are numerous. Some are
standard and some are part of Keyword's custom offering. The standard, noncustom
products are

ASCII
Asterix
Cliq Word
DEC DDIF
WPS-Plus
EDCDIC
HP Word/PC
IBM 1403 Line Printer
IBM DCA-Final Form Text
IBM DCA- Revisable Form Text
IBM DCF Script
IBM Display Write 2/3/4/5
Lotus Ami Pro
MASS-11
Microsoft Rich Text Format

Microsoft Word for Macintosh
Microsoft Word for DOS/Windows
Microsoft Word for Windows
MultiMate
ODA Q111/Q112
Picture Conversion
PostScript Writer
Quadratron Q-One
Uniplex II Plus
Wang OIS/VS Comm Format
Wang WITA
WordMARC
WordPerfect for DOS/Windows
WordStar

Conversion Categories

Keyword divides conversion features into 13 categories, which help apply a logical structure to the document conversion process. These categories are summarized here, and they represent important areas to address when choosing a document conversion product. The categories list different document conversion attributes and requirements which can then be used as a framework to generate a document conversion Request For Proposal (RFP) or Request For Information (RFI).

DOCUMENT INFORMATION   Document information applies to the entire document and includes conversion areas such as title, subject, comments, key words, document date/time, creation date/time, revision history, organization, operator, authors, attachments, language, default body style, default header style, default footer style, default note style, and default tabs. Many of these items are familiar to users of word processor applications such as WordPerfect or Microsoft Word.

SECTION INFORMATION   Section information refers to groups of pages all having the same set of format specifications and includes format constructs that handle the initial page, common page, even pages, odd pages, header definitions, and footer definitions.

PAGE LAYOUT   This category covers frames and page layout, and includes items such as page width, page height, header area, body area, footer area, print page number location, page border, and vertical centering.

FRAMES   This category covers frames, rectangular information that is part of the document but is treated differently from the surrounding text. Frame information includes items such as position, dimension, rotation, anchor, text wrap rules, transparency, borders, printing notes in a frame, repeating frames, and patterns.

TABLES   This area covers tables, which are defined as a set of rows and columns with some individual formatting. Table descriptive parameters are horizontal alignment, left offset, right offset, space before tables, space after tables, number of header rows, number of footer rows, borders, header/footer row borders, header/footer connected to body, header/footer row gap, bottom of page borders, widow count, orphan count, row properties, galley properties, cell properties, and table styles.

MULTIPLE COLUMNS   This category covers multiple columns of text on a page with description variables such as synchronized columns, vertical separation between columns, snaking columns, balanced columns, and hard column breaks.

PARAGRAPH INFORMATION   Paragraph information is specified through rulers, paragraph attributes, and styles. Descriptors in this category are indents, alignments, tab stops, leading, widow lines, orphan lines, hyphenation zone, consecutive hyphens, paragraph separation, paragraph protection, drop capitals, paragraph numbering, intercharacter white space, interword white space, paragraph pattern, paragraph alignment, paragraph borders, paragraph breaks, and line spacing.

PARAGRAPH STYLES   This category refers to a named set of properties that apply to a piece of text. Areas covered in this category are style name, next style, based on previous style, generic content, paragraph properties, and character properties.

CHARACTER INFORMATION   Character information covers character-specific styles such as boldface, italics, reverse background, elongated text, superscript, subscript, single underlines, word underlines, word and tab underlines, word and space underlines, double underlines, double word underlines, strikeouts, revision deletion, revision insertion, outlines, shadows, overbars, blinking, all caps, small caps, no caps, mixed caps, each word caps, foreground color, background color, multicharacter stacking, font name, font family, font spacing, and point size.

NOTES AND ANNOTATIONS   This area covers notes and annotation parameters such as number format, number setting, renumbering strategies, separators, text, gap measurement, leading text, continuation notices, continuation separators, and annotations.

FIELDS   This category covers miscellaneous attributes and functionalities such as style references, number of characters in a file, number of words in a file, number of pages in a file, quotes, fill-in bookmarks, fill-in prompts, one time fill-in prompts, title, subject, comments, key words, document date and time, creation date and time, organization, operator, authors, attachment files, language, and include files.

GRAPHICS   This category describes two areas of embedded graphics in the word processing file:

- Vector, Raster, or EPS Picture Class
- WordPerfect format, Encapsulated Postscript, Tag Image File, Graphics Interchange, PC Paint, MacPaint file, ITU Group 3 one-dimensional fax, ITU Group 4, Windows Bitmap, CGM,Windows Metafile Version 3, Mac PICT and AmiDraw picture type formats, internal picture capability, external picture capability, cropping, scale to frame, rotation, and inverted images

MISCELLANEOUS   Miscellaneous covers those items not covered by the other categories, such as advanced commands, auto-numbering, bookmark references, bookmarks, break types, captions, nonwrapping spaces, indexing, invisible hyphens, library documents, line breaks, line numbering, lists, sublists, primary merge lists, secondary merge lists, nonprinting text, paragraph endings, print date and time, print page number, table of contents, and table of summaries.

You can find out more information about specific products by contacting the individual vendors.

Summary

Document conversion is a complex task for which only a few vendors offer products. The list of items necessary to handle document conversion discussed in this chapter can be easily used as a basis for comparison of the various document conversion products.

Chapter Seventeen

What You Should Know About E-mail Security

E-mail security depends on all e-mail components—the DUA, MUA, DSA, MTA, and MS—functioning at the stated level of security. In other words, the security failure of one component causes the entire network to fail. In an environment where different organizations—private, public, and commercial—manage the e-mail network, total security is impossible. Some types of security are easier to implement and manage than others. For example, protecting the disclosure of your message contents is much easier than ensuring that the message is not received in another environment and read or copied there. A user may encrypt the message so that only the intended destination user may read it. However, if an MTA is not appropriately managed, the message could be copied by an intruder as the message transits the network. The message is not readable by the intruder, but could be routed to another destination or deleted.

End-to-end security deals with the source and destination e-mail systems, and *infrastructure security* with all e-mail components functioning securely. End-to-end security is attainable and is the easiest to implement. Infrastructure security, on the other hand, is not economically attainable today on any large or global e-mail network.

E-mail Security Services

The X.800 standard, called *Data Communications Networks: Open Systems Interconnection (OSI): Security, Structure, and Applications,* and the X.400 standard discuss security threats and services to combat them. The concepts discussed in these documents are generally applicable to all systems. E-mail security involves two general areas: access control and message security services.

Access Control

Electronic messaging systems can be used only by those subscribers who are known to the system, or *authenticated* users, and those who are permitted to gain entrance to the system, or *authorized* users. Access control comes into play when a user logs on to a computer system. Once the account number and password have been given, and the user is authenticated, a user is granted certain predefined rights, privileges, and services. Some users have more privileges than others. For example, system administrators have the complete set of access privileges, while an external user of the system has an extremely limited subset. Access control can be managed with either passwords or tokens. Tokens are derived from public and private key encryption technologies and are more secure than the historic password techniques.

The e-mail security services currently offered by SNADS-, SMTP-, X.400-, and NetWare Global MHS-based systems support the password method of access control. The widespread use of token access control awaits implementation on distributed, globally accessible directories. Directories are necessary to manage the keys. A user's directory entry contains a field for the user's public encryption key. This key is easily obtainable by anyone authorized to access the directory. The messages are encrypted using the public key and decrypted using the private key. Those who administer and manage the private keys are known only to the individual user. This encryption technique is asymmetric, since it uses two different keys. Rivest, Shamir, Adelman (RSA) is an example of this type of encryption. Directory management must be trustworthy to ensure key assignments are appropriate and protected. Encryption technology has been around since the late 1970s. Full scale implementation awaits the widespread deployment of X.500 directory security services and trusted organizations to manage it.

Message Security Services

Message security services protect messages from being copied, disclosed, and modified. They ensure that the source and destination addressees are who they say they are. All of these functions can be implemented using public- and private-key encryption techniques, such as RSA; and single-key encryption, such as Data Encryption Standard (DES) technologies. DES is a symmetric encryption system. Using this technique, both parties to an encryption have the same key. The originator uses the key to encrypt the document or file, and the recipient uses the same key to decrypt the file or document. Therefore, both users must maintain strict confidentiality of the key.

Overall, the security services offered by the X.400 standard cover the needs of the electronic messaging world. These services, if used on a transaction-by-transaction basis, allow:

- The sender to know that the same information that was sent was received and that the receiver acknowledged the transaction
- The recipient to know that the information received corresponds with what was sent and that the sender received an acknowledgment of such
- That the contents were not disclosed or modified during transmission
- Each party to verify what the other sent so that the other party cannot deny sending or receiving the message

These message protection services are implemented between the user's MUAs. In some instances the MTAs are involved, but most of the load falls on the MUA, which assumes the user invokes the services at each message submission.

Most of the security services described here assume the implementation of two technologies: global directories and public-key encryption.

The use of public-key encryption assumes an accessible global directory, such as X.500, to distribute the public portion of the encryption keys. These keys can be used to implement digital signatures and ensure document integrity and confidentiality. These are described in the X.509 standard entitled "The Directory - Authentication Framework."

The 1988/1992 X.400 standard addresses several security services. They fall into two general classes: those dependent on the end users and independent of the electronic messaging infrastructure, and those dependent on the electronic messaging infrastructure and independent of the end users.

The two primary areas of security services offered by the X.400 standard are described in Tables 17-1 and 17-2.

User Security Services	Description
Message authentication	This verifies the identity of the message originator.
Report origin	This verifies the identity of the delivery/nondelivery message.
Content integrity	This verifies for the recipient that the contents of the message were not modified.
Content confidentiality	This encrypts the message so it is unreadable to anyone other than the recipient and the sender.
Nonrepudiation of origin	This service gives the recipient proof of contents and proof that the message was sent by the indicated source.
Nonrepudiation delivery	This gives the originator proof of the delivery of contents to the recipient.

Table 17-1.   *End User-to-End User-Oriented Security Services*

Infrastructure Security Services	Description
Security labeling	This supports the categorizing of a message, with respect to sensitivity.
Nonrepudiation of submission	This gives the originator proof that the message was submitted to the intended recipient.
Flow confidentiality	This supports concealment of the message flow through MTAs.
Sequence integrity	This provides the recipient proof that the sequence of messages was not changed.
Proof of delivery	This verifies for the originator that the contents were received by the actual recipient.
Proof of submission	This verifies for the sender that the message was submitted to the VAN.
Secure access management	This sets up security and identification between components of the delivery system.

Table 17-2.   *Infrastructure-Oriented Security Services*

Summary

The X.400 standard discusses a vigorous and economical set of security services covering a broad spectrum of e-mail systems and implementations. These services meet the security needs of the business world. Though they define categories applicable to all e-mail systems, some are not currently implemented because they await the wide scale implementation of the global directory.

Chapter Eighteen

Electronic Messaging APIs

A n application program interface (API) is a set of standard subroutines or function calls that provides the connection between an application and an end user's e-mail system. APIs assist developers in creating platform-independent message user agents (MUAs) and in creating applications that are e-mail enabled or e-mail aware. Four major multivendor messaging APIs now exist. They are Vendor Independent Messaging (VIM), lead by Lotus; Common Messaging Calls (CMC), lead by the X.400 API Association; Messaging Application Program Interface (MAPI), lead by Microsoft; and XMHS and XMS APIs by X/OPEN. This chapter describes the first three of those platforms and gives examples of the various API styles and capabilities.

Through the use of APIs, an MUA such as cc:Mail can be easily transported to Novell, Banyan, LAN Manager, and LAN Server environments without the need for recoding. APIs also allow different MUAs to share the same message store (MS) for file storage and retrieval, which makes the system easier to manage. The widespread use of messaging APIs cuts costs and increases the user's choice of MUAs over a wide range of messaging infrastructure platforms.

Overview

An API implements a standard way to access and submit requests for services from the MUA. All three of the APIs discussed in this chapter help users submit, retrieve, address, and query message systems for addresses and status. The following is an example of an API function call for creating a message from the VIM API suite:

VIMCreateMessage (*Session, psType, pMessage*)

The variables in this example represent the following:

- *Session* is the identification for this VIM session
- *psType* specifies the type of message being composed (VIM_MAIL is one of the values)
- *pMessage* points to a VIMMsg variable that receives the identifier of the created message

Vendor Independent Messaging (VIM)

VIM is an industry standard API that enables developers to write applications that operate on multiple e-mail messaging systems. The VIM specification was developed by Lotus and is supported by Apple, Borland, IBM, MCI, Novell, Oracle, and WordPerfect.

The VIM API allows a developer to create an application that can perform the following:

- Compose and send messages
- Receive, read, store, and file messages

- Simple mail sending
- Manage address books

Depending on the underlying e-mail system, a message being composed could contain text, file attachments, images, and facsimiles.

VIM Message Structure

Several different types of messages are possible. VIM supports messages relating to e-mail, EDI, and calendaring. The e-mail message, specified by VIM_Mail, consists of a heading with attributes such as the recipient and originator address, time, date, priority, and message class. The body of the message may be a Note Part, file attachment, or application defined item.

NOTE PART   A Note Part is the actual message body in Lotus Notes terminology. Note Parts can include unformatted text strings, rich text, fax images, Microsoft DIB color bitmap images, Apple PICT files, Microsoft metafiles, sound, and video.

FILE ATTACHMENTS   File attachments are also message components. Attachments, such as WordPerfect or Microsoft Word documents, for example, are like Note Parts but are not well defined in VIM. The application that creates the file defines the attachment format.

APPLICATION-DEFINED ITEMS   Application-defined items allow VIM to specify additional types for future expansion, such as holograms, animated graphic images, and so on.

Data Attribute Encoding

VIM uses a data structure logically similar to ASN.1 encoding. Each item is composed of three parts: attribute, size in bytes of buffer, and buffer. This logical structure reduces the complexity in designing the computer programming code when using the VIM API. See Chapter 12, "X.400 Interpersonal Messaging," for a description of ASN.1 encoding.

Character Sets

VIM supports several character sets, including IBM code, Microsoft code, Lotus Multibyte Character Set, Unicode, and Apple International String.

VIM Simple Message Interface (SMI)

VIM supports a full function API and a Simple Message Interface (SMI) API that is easier to use than the full function VIM API. Four self-explanatory calls are defined for the SMI API set. They are *SMISessionsSetup*, *SMISendDocuments*, *SMISendMail*, and

SMIAddressAndSend. Each of these has a parameter list to convey the message information.

VIM Message Creation Process

Four VIM function calls (*VIMCreateMessage, VIMMessageHeader, VIMSetMessageRecipient*, and *VIMSetMessageItem*) are used here to show an example of creating a message using the VIM API suite. However, the sequence logic is generic in that the MAPI and CMC APIs discussed later in this chapter also use the same general sequence to create a message.

The following list is a step-by-step example of how VIM is used to create a message.

1. First, a session is opened with the underlying message system and a unique session ID is assigned.

2. *VIMCreateMessage* is called, referencing the session ID, with VIM_MAIL set to indicate this is an interpersonal e-mail message and not some other type of message. This call has a pointer to the buffer, which is used to construct the message heading and body.

3. *VIMMessageHeader* is called several times, each time setting one more of the message header attributes as specified in the VIMMessageHeader function call, described later in this section.

4. *VIMSetMessageRecipient* is used several times to setup all the To, CC, and BCC fields in the message header.

5. *VIMSetMessageItem* is called zero several times to attach files or construct the Note Part body of the message, although a message does not have to have a body.

6. The message is then closed or sent using the *VIMSendMessage* or *VIMCloseMessage* function calls.

VIMCreateMessage

The initial message template is produced by using the Create Message function call, *VIMCreateMessage(Session, psType, pMessage)*. *Session* is the session identification for this VIM session. The *psType* specifies the type of message being composed. VIM_MAIL is one of the values. The *pMessage* points to a vimMsg variable that receives the identifier of the created message.

VIMSetMessageHeader

The usual message heading information is entered into the message with the Set Message Header function call, *VIMSetMessageHeader(Message, selAttr, AttSize, Patt)*. Here, *Message* is the message identifier. The *selAttr* type has several possible values as described in Table 18-1. *AttSize* is the size of the message part, and *Patt* is a pointer to the attribute descriptor block.

VIMSetMessageRecipient

A recipient is entered in the message header by using the Set Message Recipient function. Each call enters one recipient in the message header. The function is stated *VIMSetMessageRecipient(Message, Selector Class, Recipient). Message* is the message ID. *Selector Class* designates To, CC, or BCC. *Recipient* points to one of two values, entity or distribution list.

VIMSetMessageItem

The Set Message Item function call defines the body parts and attachments of the message. The call is stated *VIMSetMessageItem(Message, Class, Type, Flags, Name, Description). Message* is the message ID. *Class* points to one of three values: Note Part, Attachment, or Application defined, which were described previously. *Type* points to

Value	Description
SUBJECT	The pointer to the subject string
PRIORITY	The pointer to one of three values: LOW, NORMAL, or HIGH
DELIVERY_REPORT	A pointer to one of two values: YES or NO
NONDELIVER_REPORT	A pointer to one of two values: YES or NO
NONDELIVERY_CONTENTS	A pointer to one of two values: YES or NO
ENCRYPT	A pointer to one of two values: YES or NO
ENCRYPT_WITH_KEY	A pointer to a unique encryption key for the message
EXPIRATION_DATE	A pointer to the date the message expires
SENSITIVITY	A pointer to one of four values: NORMAL, PRIVATE, PERSONAL, or CONFIDENTIAL
IN_REPLY_TO	A pointer to the unique message ID for which this message is a reply
RESPOND_BY	This is the date by which the recipient should reply
KEYWORD	Pointer to a keyword for a message; not all systems support this feature
RETURN_RECEIPT	A pointer to one of two values: YES or NO
SAVE	A pointer to one of two values: automatically save on receipt or don't automatically save on receipt
DRAFT	A pointer to one of two values: draft message or not a draft message

Table 18-1. *Values for the VIM selAttr Parameter*

one of eight different values in a Note Part class: Text, Unwrapped text, Microsoft Rich Text, Macintosh Styled text, PCX Fax format, Macintosh PICT format, QuickTime sound and video, or Macintosh Fax image. The *Flags* field indicates a series of bit flags. *Name* is the title of Note Part, and *Description* is the description of the contents of the attachment or Note Part body part.

VIM Address Format

The *vimAddress* is not a function call, but a definition of the address data. The address is composed of two parts. The first part defines the type of address and the second part is the address string. The address type has values such as: cc:Mail, MHS, MSGMGR, Notes, OCE, or X.400. The address string can contain up to 1,024 bytes.

VIM Summary

The VIM API is a powerful programmer's interface for developing applications that can be used on e-mail. Over 60 different calls exist in the VIM API set and the few presented above give an example of the formats, styles, and capabilities. The additional VIM functions cover areas such as Session Management, Message Creation and Submission Functions, Message Container Functions, Message Access and Attribute Functions, and Address Book Functions.

Common Messaging Calls (CMC)

The X.400 API Association's Common Messaging Calls (CMC) version 1.0 was released in June 1993. The X.400 API Association (XAPIA) agreed to establish a vendor-independent set of messaging APIs in 1992 due to the lack of cooperation between Lotus and Microsoft; Microsoft APIs did not support VIM, and Lotus did not support MAPI. Developers were therefore spending major resources to support both sets of APIs. The members of the X.400 API Association are

CTC	Digital Equipment Corportion
RAM Mobile Data	OSIWare
First Telecom	Bull
AT&T	Retix
Microsoft	ISOCOR
Datalogic	IBM
Action Consulting	Novell
Iris Associates	cc:Mail/Lotus
Tandem Computer, Inc	

The CMC API set supports four main functions and is currently among the simplest to implement. While the API functions are easy to implement, they only support the major e-mail functions. More sophisticated applications require the use of the MAPI or the VIM API sets. The four functions supported by CMC—sending messages, reading messages, address lookup, and administration—are described next.

Sending Messages

The sending of messages is supported by two API function calls, *Send* and *Send Documents*. *Send* is a function call used to send an e-mail message. *Send Documents* is a more limited function call used to send an e-mail message, based on macros such as those used within spreadsheet programs.

Receiving Messages

The receiving and reading of messages is supported by three function calls: *ActOn*, *List*, and *Read*. *ActOn* performs an action on a specified message. *List* contains summary information about messages that meet specified criteria. *Read* permits the API to read the message.

Looking Up Names

The address lookup area has one API call, *LookUp*, which is used to search for addressing information.

Administration

The administration of the underlying e-mail system services requires four API calls: *Free, Logon, Logoff,* and *Query. Free* is used to free memory after other calls are complete. *Logon* is used to establish the session with the underlying messaging services, and *Logoff* is used to terminate it. The last call, *Query*, is used to determine information about supported CMC services.

CMC Messaging Calls

APIs are how a programmer sees the message structure and underlying messaging services. A programmer using an API does not need to know as much about the underlying messaging systems or details about the particular e-mail system. The underlying messaging system will transparently convert the message structure to its own type for the programmer. The structure might be converted to a cc:Mail or a QuickMail message format by the receiving routine. After a message is received and read by the CMC routines, it is converted back to the e-mail system's format.

The best way to get a feel for what these APIs are all about is to give some examples of the API function calls themselves. Several of the messaging calls are described next. These examples show the parameters necessary to implement the call and the structure of the call. Don't get buried in the details; just peruse the calls and get a feeling for CMC's style and capabilities as compared to those of VIM and MAPI.

Message Definition

The message structure must be defined within CMC showing the parts of the CMC message. Remember, the message format will be converted to the underlying system format transparently to the applications interface programmer. The syntax for the call is

MessageStructure::= *MessageReference, MessageType, Subject, TimeSent, TextNote, Recipient, Attachments, MessageFlags, TextNoteHandling*

Message Reference is the unique message ID. *Message Type* has several possible values: Interpersonal, Interpersonal Receipt notification, Interpersonal Nonreceipt notification, Delivery report, and Nondelivery report. Others may be defined by bilateral agreement between the participating application and the e-mail system. *Subject* is the topic of the message. *TimeSent* is the time and date the message was submitted to the message handling system. *TextNote* is the normal message text. *Recipient* is the pointer to the array of the recipients of the message. *Attachments* is the pointer to the array of the attachments to the message. *MessageFlags* are bit flags which have the following values: Read/not read, Text-note handling, sent/unsent, or last/not last element in the array.

Address Definition

The CMC electronic messaging address is composed of six parameters in the function lists. The address syntax is

Name::= *Name, Name_Type, Address, Role, Recip_Flags, Recip_Extentions*

Name is the name to be displayed and *Name_Type* has one of three values: unknown, individual, or distribution list. *Address* is the e-mail address as required by the underlying messaging system. *Role* specifies the type of addressee with values of: To, CC, BCC, Originator, or Authorizing User. *Recip_Flags* are bit flags per each recipient with values of: ignore, truncated, and last_element. *Recip_Extentions* point to the first element in an array of per recipient extensions.

Send Function

The definition of the Send function call is

Send(*Session_ID, Message, Send_flags, UI Identifier, Send_extentions*)

Session_ID is the identification of the session. *Message* is the same message structure defined previously in this section. *Send_flags* are a series of bit flags. *UI Identifier* is a user interface indicator and is not always supported by all implementations. *Send_extensions* is the pointer to the extension array.

Act On Function

The Act On function is used to manage the messages with such calls as delete or store, for example. The call is formatted in this way:

> *ActOn(Session ID, Message Reference, Operation, Act_On_Flags, ui_id,*
> *Act_on_extensions)*

Session ID is the same as defined previously. *Message Reference* is the reference number of the message to access. *Operation* has two values: look in list of extensions for the action or delete. *Act_on_flags* is a bit mask. The *ui_id* parameter is a user interface specifier, for example, a dialog window that would appear on the user's e-mail application screen. *Act_on_extensions* is a pointer to the array of extensions.

List Function

List allows a user to list the messages located in the message store. The call is stated in the following format:

> *List(Session_id, Message_Type, List_flags, Seed, Count, ui_id, Message_summary,*
> *Extensions)*

Session_id has been defined previously. *Message_type* is the returned information on the message. *List_flags* is a bit flag with four values: allow error dialog box on error, unread message only, only message reference field valid, and count of messages meeting specifications. *Seed* is the message reference to use in a search function. *Count* is the maximum number of messages to return to the requester. *ui_id* has been defined previously. *Message_summary* is the address of the array holding the result of the request. *Extensions* is the pointer to the extension array.

Read Function

The Read function is used by the API to actually read the message. The call is stated thus:

> *Read(Session Id, Message Reference, Read flags, Message, Ui Id, Read_extensions)*

Session_id, *Message_reference*, *Message*, and *Ui_id* have been defined previously. *Read_extensions* is an array of read extensions.

CMC Summary

CMC calls are usable in Windows as well as other platforms, unlike MAPI calls, which are only usable in Windows systems. These are the simplest of the current messaging APIs to implement. Also, since both Microsoft and Lotus belong to the X.400 API Association, they support the Windows and the non-Windows environments equally well.

Messaging Application Program Interface (MAPI)

Simple MAPI is a Windows messaging API with several functions. An example of how a message is created and sent is given in the VIM section. The specifics are not the same, but the logical flow of how the API calls are made is. The Simple MAPI function calls are described in Table 18-2.

Three data structures are also defined for the MAPI interface. They are *MapiFileDesc* which contains file attachment information, *MapiMessage* which contains message information, and *MapiRecipDesc* containing recipient information.

The *MapiMessage* data structure has pointers to the *MapiFileDesc* and the *MapiRecipDesc* structures. The parameters Originator, Recips and Files in the MapiMessage data structure point to the MapiRecipDesc, the MapiRecipDesc and MapiFileDesc respectively.

MAPI Function Call	Description
MAPILogon	Start messaging system session
MAPIFindNext	Return ID of next mail message of specified type
MAPIReadMail	Read a mail message
MAPISaveMail	Save a mail message
MAPIDeleteMail	Delete a mail message
MAPISendMail	Send mail message
MAPISendDocuments	Send mail—not as functional as SendMail
MAPIAddress	Address a mail message
MAPIResolveName	Dialog box prompt for more information
MAPIDetail	Dialog box with recipient detail
MAPIFreeBuffer	Free memory
MAPILogoff	End messaging system session

Table 18-2.   *Simple MAPI Function Calls*

A few of the MAPI calls are described next to show the overall architecture of the calls. Several more calls exist that are not described herein.

MAPIAddress Function

MAPIAddress is used to create or modify a set of address entries. This generates a Windows dialogue box for input and address modification. The call may be formatted as follows:

> *MAPIAddress(Session, UIPara, Caption, EditFields, Labels, nRecips, lpRecips, Flags, Reserved, NewRecips, lpNewRecips)*

Session and *Flags* have been defined previously. *UIPara* is the parent window handle for the dialogue box. *Caption* is the title for the dialogue box. *EditFields* and *Labels* are used in designing the dialogue box. The recipients of a message are quantified by *nRecips* and listed by *lpRecips*. *Reserved* indicates that this field is reserved for future use. *NewRecips* is the pointer to the number of new recipients and *lpNewRecips* is a pointer to the final array of recipients.

MAPISendMail Function

MAPISendMail sends a mail message with optional recipients, subject, file attachments and message text. The call is formatted as follows:

> *MAPISendMail(Session, UIParam, Message, Flags, Reserved)*

Session, Message, Flags, and *Reserved* have the same definition as in the previous sections. *UIParam* is the handle of the parent window for the dialog box.

MapiMessage Structure Definition

MapiMessage defines the data structure for the message. This data structure is referenced by function calls such as: *MAPISendMail*. The structure of the MAPI message is defined in Table 18-3.

MapiRecipDesc Definition

This structure, *MapiRecipDesc*, contains the recipient addressing information. It is pointed to by the MapiMessage structure. *Reserved* has been previously defined. *RecipClass* contains one of four values: To, CC, BCC, or ORIG. *Name* is the displayed name of the message recipient. *Address* is the message specific address data. Both *EIDSize* and *EntryID* are message system specific values.

MapiFileDesc Definition

MapiFileDesc is used to describe the file attachments associated with a message. It is pointed to by the MapiMessage structure.

MAPI Message Structure	Description
Subject	Subject text
NoteText	A pointer to message text string
MessageType	The pointer to one of several values: Null means Interpersonal Message(IPM); other values mean other types of messages such as EDI or calendaring
DateReceived	The pointer to the date data string in the format: YYYY/MM/DD HH:MM
ConversationID	The pointer to the string containing the conversation thread for this message
Flags	A bit mask with several values: unread, receipt requested, and sent
Originator	The pointer to *MapiRecipDesc* structure describing the originator of the message; *MapiFileDesc* is described elsewhere in this section
RecipCount	The number of recipients
Recips	A pointer to array *MapiRecipDesc* of recipients; *MapiRecipDesc* is described elsewhere in this section
FileCount	The number of *MapiFileDesc* arrays pointed to by Files
Files	The pointer to the array of file attachments

Table 18-3. *MAPI Message Structure Definition*

Reserved has been previously defined. *Flag*, in this case, describes the embedded or attached OLE object. *Position* is the offset in the message where the attachment should be inserted. *PathName* is the pointer to full pathname of the attachment file. *FileName* is the pointer to the filename as seen by the message recipients, and *FileType* is the type of file.

MAPI Summary

MAPI is Microsoft's messaging API specifically developed for the Windows environment. To appreciate the richness of the MAPI calls, one must read the MAPI manuals. The previous section provides a better understanding of the MAPI structures, formats, styles, and capabilities.

Summary

All of the APIs discussed in this chapter are similar in logic, with CMC being a vendor-neutral messaging API. Many third parties are supporting these. Whether any one of them will dominate the marketplace is hard to say. However, since MAPI is for Windows specifically, and Windows is the leading PC operating system, MAPI holds the lead.

Appendix A

Value Added Network RFP/RFI Technical Skeleton

This appendix presents a sample Request For Proposal/Request For Information (RFP/RFI). RFP/RFIs are used to gather technical specifications, cost information, and administrative data about a specific vendor's product(s) or service(s). *RFIs* are used to gather basic information to limit the number of vendors considered or to find out the current state of a product or service in the industry. *RFPs* are usually more formal, issued by the organization's procurement department, and are used when you purchase a product or service. The RFP/RFI presented in this appendix is included to help you gather the information necessary to select a value added network (VAN). Much of the appendix can be taken as is, but some will require decision making on your part. In any event, if you want to make a decision on a VAN, you might start by skimming this appendix, then adapting the material here for your own RFI and submitting it to your prospective providers.

Technical

1. COMMUNICATIONS

1.1 Provide VAN trading partners communications access that is flexible enough to accommodate a high degree of sophistication. At a minimum:

1.1.1 X.PC or Xmodem with CRC protocol, with line speeds between 2400 bps and 9600 bps for asynchronous users

1.1.2 Line speeds of 2400 bps, 4800 bps, and 9600 bps for synchronous users of the network, and support for X.25 packet switching

1.2 Support X.25 to send/receive data upon request.

1.3 Support dial-out and send/receive data on a time schedule. Sessions will take place on a specified time schedule or as necessary depending on the urgency of the message.

1.4 Provide network access engineered to the P.01 grade of service or better.

1.5 Provide network and related systems availability 24 hours a day, seven days a week.

1.6 Provide international access, including local support, to western Europe, the Far East, South America, and Israel, either directly or through an interface with other networks or telecommunications companies.

1.7 Provide access to the network via a dial-up local number and/or dial-up toll-free number.

1.8 Provide capability to send and receive messages within the same communications session in a store-and-forward batch mode.

1.9 Provide error-free communications within the network and during dial-up connections.

1.10 Have local access in population areas in excess of 100,000. It is also desirable that there be local access in all 50 states and the District of Columbia. (*Local access* means no telecommunications charge; therefore an 800 number would be an acceptable access method).

1.11 Support transmission of binary data.

1.12 Provide routing restriction capability.

Questions

1.13 What are the protocols you can support in a personal computer environment with line speeds between 300 bps and 9600 bps for asynchronous users? Define them.

1.14 Does the network support interactive transmission or only batch file transfer?

1.15 What is the network availability? Reliability? How do you demonstrate same? Please provide availability statistics for the last 24 months.

1.16 Does the network interface with proprietary and closed networks?

1.17 Does the network provide for network standards translation capabilities (for example, X.400-to-non-X.400 trading partners)? Where is the translation service performed? Is the translation software part of the standard offering?

1.18 What are the numbers and locations of access points to the network?

1.19 How are binary files supported? What types of binary files are supported? How are graphics handled?

1.20 How are messages that contain compound documents handled (for example, text, graphics, and multiple binary attachments of differing types, such as ODA and ODIF standards?

1.21 How is the routing of messages restricted?

1.22 How will the connection be made to ensure it is not an unrestricted session-to-session connection, and how will it be secured?

1.23 What are the international sites you support? Provide a list.

2. INTERCONNECTION WITH OTHER NETWORKS

2.1 Transmissions requiring interconnection with other networks should occur once every 30 minutes or less.

2.2 Interconnect with any/all networks (for example, GEIS, MCI, IBM, US Sprint, AT&T, Western Union, or CompuServe) we require in order to communicate with trading partners, or have plans in place with these networks to provide such services in the immediate future.

2.3 Interconnecting with other networks should be transparent and not require any additional or special processing on the part of us or our trading partners.

2.4 Provide positive methods for assuring that all messages are received by and from your network and other networks.

2.5 Provide capability to interconnect our VAN service with the VAN services of various trading partners. This service should either currently support X.400 (1988 or 1992 versions) or Internet, or have a plan in place to support same in the future.

Questions

2.6 What assurance is there of mailbox-to-mailbox deliveries and receipts to other VANs?

2.7 What is the status of interconnections with other public service providers: (1) in production, (2) in testing, (3) announced only, or (4) not planned?

2.8 What is the expected e-mail delivery time for trading partners on the VAN? What is the expected e-mail delivery time for trading partners on other interconnected services?

2.9 What are existing and future network plans as outlined in VII.2.2.2? Describe them and provide details.

3. STANDARDS

3.1   Provide support for ITU X.400 (1988) and transition to X.400 (1992) standards as they become available.

3.2   Support the ITU X.500 (1988/1993) standards.

Questions

3.3   What 1988 and 1992 X.400 standards will be supported?

3.4   What is the status of any X.400 EDI (EDIFACT, X12) offering?

3.5   What is the status of any X.500 efforts: (1) in production, (2) in testing, (3) announced only, or (4) not planned?

4. GATEWAYS

4.1   What types of gateways are supported (for example, Internet, X.400 (1988), SNADS, MHS, cc:Mail, Notes, Microsoft Mail, and others)?

4.2   To what degree are supported gateways certified: in production, in testing, and so on?

5. BACKUP AND RECOVERY

5.1   The VAN should protect and automatically resend a message in the event of a line disconnect.

5.2   A documented disaster recovery plan that is tested at least four times per year should be in place. The plan should include provisions to provide us with VAN network services within 24 hours following the onset of a network outage.

5.3   Is there system and network redundancy? If so, describe.

6. SECURITY

Questions

6.1   Have in place appropriate and proven access security measures such as:

    6.1.1   Provisions for password maintenance and validation/authentication

    6.1.2   Automatic termination or suspense of access after repeated password violation

    6.1.3   Confidentiality of passwords between the network and trading partners

    6.1.4   Knowledge of passwords within networks limited to key personnel on a need-to-know basis

6.2   Agree that you will not share with any other organization information proprietary to us and/or our trading partners. This prohibits the use of company names in publications, partnership arrangements, transaction contents, administrative and security information, and so on, without our express written approval.

6.3   Provide the capability to handle ASCII and binary-encrypted messages. Describe how this works from the gateway and your VANS.

Questions

6.4   How is data controlled and protected?

6.5   How will encrypted data, if provided or planned for in the future, be handled?

6.6   How will routing be done to preclude the use of microwave and satellite transmissions of encrypted messages? Provide information detailing this.

6.7   How does the network ensure that no other sender or other part of the firm gains access to our messages?

6.8   How would you recommend we activate a disaster recovery plan if there is a loss of a major location?

6.9   What security features are available?

7. PERSONAL COMPUTER SERVICES

7.1   Support access and usage of VAN services via a personal computer.

7.2   Provide the following personal computer capabilities, mainly as a potential VAN solution for our trading partners.

7.2.1   Entering and editing of messages directly on the personal computer

7.2.2   Automatic local translation to ITU X.400 or Internet

7.2.3   Transmission and receipt of messages

7.2.4   Automatic generation of acknowledgements

7.2.5   Connection to the VAN network

7.2.6   Quick and easy implementation of the system

7.2.7   Front end to computer mainframes or operation as a stand-alone personal computer

7.2.8   Directory services for PC-based end nodes

7.2.9   Support for remote e-mail packages from QuickMail, cc:Mail, Microsoft Mail, and so on

Questions

7.3   Are there DOS, Windows, UNIX, and Macintosh-based user agents available?

7.4   Do the user agents allow the creation/storage of messages on the user device?

Delivery Forms Supported

Non-X.400 protocols in the following could be either Internet, SNADS, MHS, or other proprietary protocols important to your company.

1.   Support X.400-to-X.400 messages to the same VAN.
2.   Support X.400-to-non-X.400 messages to the same VAN.
3.   Support X.400-to-X.400 messages on different VANS.
4.   Support X.400-to-non-X.400 messages on different VANS.
5.   Support X.400-to fax messaging.
6.   Support X.400-to-postal mail messaging.
7.   Support X.400-to-international postal mail messaging.
8.   Support X.400-to-TWX messaging.
9.   Support X.400-to-telex messaging.
10. Support X.400-to-overnight mail messaging.
11. Support X.400-to-courier messaging.
12. Support letterhead capability.
13. Support signature capability.
14. Support X.400-to-international connections.
15. Support return receipts and delivery receipts.

Questions

16. Are return receipts automatic or optional?
17. Are directories available to all users on the same VAN?
18. What is available to simplify X.400 and proprietary addressing for the user?
19. What administrative functions are required to support the various messaging functions?
20. Are fax cover pages included in the page count?

Administration

1. Provide audit/status reports and data file provisions to trace all activities that have occurred between all parties, networks, and trading partners. This information should be available to us within two hours following transmission and be maintained for a period of 60 days.
2. Provide status reports containing information pertaining to undelivered and/or unacknowledged transactions.
3. Provide billing/cost reports and data file provisions that detail transaction set charges by user, day, peak times, off-peak times, and so on.
4. Provide historical summary reports indicating chronological sequence of transaction events and any or all related information. This information should be available for a minimum of 60 days.
5. Provide historical reporting provisions of activity usage to assist us in monitoring and managing VAN traffic.
6. Provide general capability to obtain, in both batch and online mode, data required for us to effectively manage, operate, and control the VAN process.
7. Provide billing options flexibility to include electronic delivery of billing data (direct or magnetic tape) and allow for any distribution of charges by divisions, departments, projects, and users.
8. Provide consolidation of multiple accounts with the network in order for us to take advantage of volume discounts.
9. Provide for easy registration of users, routing control, and access to X.400.
10. Provide routing and control mechanisms that limit access and use of VAN services.
11. Provide for directory services.

Questions

12. Is user administration (setup and maintenance) accomplished online, batch, or both?
13. Who is responsible for registering user accounts?
14. Is there a formal checklist for adding users to a network? If so, attach a copy.
15. What is the average length of time required to add users to the network?
16. How is access to non-X.400 users accomplished?
17. Do users have to be registered with the VAN before they will forward e-mail to our gateway?
18. Can inbound messages be restricted from delivery to our gateway?
19. How long are messages stored before an error is generated? How long are they kept until they are deleted? What is the cost of message storage on a per message basis?
20. Is an administrative mailbox provided?

21. What categories of mailboxes are available?
22. Who administers the mailboxes?
23. How are directory services provided? How are directories populated? How is access provided to the user? How is security on directories provided?
24. What administration procedures are employed? Describe them and include management reports, billing, and charge-back to users.
25. How long does the VAN attempt to deliver the message before notifying the originator that the message is undeliverable? Can the number of tries be specified by the user?
26. Does the VAN specify to the originator why the message was undeliverable?

Installation

1. Provide training, consultation, and documentation support for the installation, testing, and use of the network as well as the hardware and software acquired from your company. This should include technical, operational, and administrative support, and so on.
2. Provide documentation in a format that allows for easy insertion of upgrades.
3. Provide technical and user instructions and consultation beyond that which is included in standard manuals or documentation as required.
4. Provide testing arrangements that allow for sending and receiving messages prior to sending live messages.
5. Provide administrative assistance and support in the establishment of trading partner conferences and relationships.

Questions
6. What steps are entailed in the installation?
7. What types and quantity of documentation, training, and consultation are provided? Is quick reference documentation available for the end user?

Customer Support

1. Maintain adequate customer service and technical resource personnel to adapt to changed and enhanced functionality and resolve customer service problems.
2. Provide capability to resolve our network problems in a timely manner. Under normal circumstances, resolution of the problem should occur within 24 hours.
3. Provide access to a toll-free telephone number that is staffed 24 hours a day, seven days a week, by a technically qualified customer service contact with whom we can discuss and resolve problems relating to installation and use of the VAN network service.
4. Provide capability to prioritize and assign an identifier to each contact and problem for tracking and feedback purposes. Under normal circumstances, the vendor should respond to our problem, not necessarily with the solution, within one hour.
5. Be an active participant in Internet, ITU X.400, and X.500 standards development, general industry, and other VAN-related groups.
6. Maintain awareness of other suppliers, VAN products, services, and their respective utilization.

7. Participate actively in the development and implementation of further advances in VAN software products and services.

8. Sponsor user groups in order to facilitate and improve VAN-related capabilities as a result of common user group concerns and priorities.

9. Provide VAN consulting and project management resources to assist us in VAN related areas, if required, including ITU X.400, Internet, X.500, and other standards, theories, practices, and so on.

10. Provide capability to trace lost messages once they are received by the VAN.

Questions

11. Is there on-site training for users, operations, and technical personnel?

12. Is an on-site management review provided?

13. What training materials and newsletters are provided?

14. Is there a support group, and if so, where is it located?

15. What are the network expansion plans? Please provide specific details.

16. What specific restrictions relating to the network service should we be aware of?

17. What is the recommended interface with your customers (for example, direct user interfacing with the VAN, with a company liaison, and so on)?

Costs

The only way to effectively assess and predict the cost on a VAN-by-VAN basis is to create a likely usage to which clients respond. The various VAN costs can then be compared. Though this is the only way to evaluate VAN costs, it has two disadvantages: the comparison may take major manpower and the scenarios selected may not be appropriate or realistic. Whatever the case, the scenario cost comparison method is highly recommended.

1. Each bidder should submit published price lists for all services. Bidders are to submit as much information as is necessary in a clear and easily understood manner, to properly reflect all costs associated with each specific service you offer as a result of this RFP.

2. All costs associated with these types of services must be predictable and based on some specific set of algorithms, and so on. It is particularly desirable that these costs be expressed in neutral, measurable terms, such as characters transmitted or received and/or messages sent or received, or staff hours expended by labor category. We expect that there will be discounts for volume and off-peak usage of network-related services.

3. Each cost component response should include an identifier number (for example, VII.5.1, where "VII" identifies the cost response, and "5.1" identifies the specific cost component).

4. In order for us to properly understand and evaluate the costs we would incur for utilizing your company's VAN services, we require that you complete a cost estimate for various assumptions and hypothetical communications sessions.

5.   It is estimated that approximately 50,000 X.400/Internet messages, five million fax pages, 12,000 telexes, and various related messages will have been processed via our strategic VAN within the next five years. By our best estimate of the most realistic scenario of implementation and data volumes for these messages, the planned growth rate is 10 percent the first year, 15 percent the second year, and 25 percent per year the third, forth, and fifth years.

6.   The exact volumes are subject to variation depending on the ability of the selected bidder to achieve implementations and/or support various standards as well as our participation. We fully expect to implement many other forms of business communications via your VAN during the next five years; however, we have elected not to include them in this RFP.

7.   Given the assumptions and hypothetical communication sessions below, calculate the costs that would be incurred by us. Please include details of all cost components depicting the methods used to arrive at each cost.

   7.1   Assume that we access the network during normal business hours (from East Coast to West Coast, 0800-1945 hours, EST). We have users in the continental United States.

   7.2   Assume 100 messages (X.400 and/or Internet and/or proprietary) consisting of 500 characters each and that these 500 characters include all required heading and trailer records (envelopes).

   7.3   Assume 100 messages (X.400 and/or Internet and/or proprietary) consisting of 2,500 characters each and that these 2,500 characters include all required heading and trailer records (envelopes).

   7.4   Assume 100 messages (X.400 and/or Internet and/or proprietary) consisting of 7,500 characters each and that these 7,500 characters include all required heading and trailer records (envelopes).

   7.5   Assume 100 messages (X.400 and/or Internet and/or proprietary) consisting of 7,500 characters each and that these 7,500 characters include binary and textual information and all required heading and trailer records (envelopes).

   7.6   Assume 100 letters, to be delivered in the continental United States via U.S. Postal Service, consisting of three pages each.

   7.7   Assume 100 letters, three pages each, to be delivered internationally via postal services to each of the following locations: Seoul, South Korea; Frankfurt, Germany; and London, England.

   7.8   Assume 100 letters, six pages each, to be delivered overnight in the continental United States.

   7.9   Assume 100 letters, six pages each, to be delivered overnight internationally via overnight services, to each of the following locations: Seoul, South Korea; Frankfurt, Germany; and London, England.

   7.10   Assume 100 fax messages, each ½ page in length, to be delivered in the continental United States.

   7.11   Assume 100 fax messages, each 4½ pages in length, to be delivered in the continental United States.

   7.12   Assume 100 fax messages, each ½ page in length, to be delivered internationally to each of the following locations: Seoul, South Korea; Frankfurt, Germany; and London, England.

7.13   Assume 100 fax messages, each 4.5 pages in length, to be delivered internationally to each of the following locations: Seoul, South Korea; Frankfurt, Germany; and London, England.

7.14   Assume 100 telex messages, each four minutes in length, to be delivered internationally to each of the following locations: Seoul, South Korea; Frankfurt, Germany; and London, England.

7.15   Assume that we send the following via your VAN network:

7.15.1   100 (X.400 and/or Internet and/or proprietary) messages, as specified in 5.2, to users on your network or another provider's network.

7.15.2   100 (X.400 and/or Internet and/or proprietary) messages, as specified in 5.3, to users on your network or another provider's network.

7.15.3   100 (X.400 and/or Internet and/or proprietary) messages, as specified in 5.4, to users on your network or another provider's network.

7.15.4   100 (X.400 and/or Internet and/or proprietary) messages, as specified in 5.5, to users on your network or another provider's network.

7.15.5   100 letters, to be delivered in the continental United States via U.S. Postal Service, as specified in 5.6.

7.15.6   100 letters, to be delivered internationally via postal services, as specified in 5.7, to each of the following locations: Seoul, South Korea; Frankfurt, Germany; and London, England.

7.15.7   100 letters, to be delivered overnight in the continental United States, as specified in 5.8.

7.15.8   100 letters, to be delivered overnight internationally via overnight services, as specified in 5.9, to each of the following locations: Seoul, South Korea; Frankfurt, Germany; and London, England.

7.15.9   100 fax messages, to delivered in the continental United States, as specified in 5.10.

7.15.10   100 fax messages, to be delivered in the continental United States, as specified in 5.11.

7.15.11   100 fax messages, to be delivered internationally to each of the following locations: Seoul, South Korea; Frankfurt, Germany; and London, England; as specified in 5.12.

7.15.12   100 fax messages to be delivered internationally to each of the following locations: Seoul, South Korea; Frankfurt, Germany; and London, England; as specified in 5.13.

7.15.13   100 telex messages to be delivered internationally to each of the following locations: Seoul, South Korea; Frankfurt, Germany; and London, England; as specified in 5.14.

7.16   Please specify all discounts offered for larger volumes of the above.

8.   Please specify costs for each of the following:

8.1   Individual mailbox fee

8.2   Letterhead

8.3   Signature graphics

8.4   Return receipts

8.4.1   X.400 messages

8.4.2   Fax
8.4.3   Telex
8.4.4   Postal
8.4.5   Courier

9. Provide separate prices that reflect network access engineered to P.01, P.05, P.10, and P.25 grades of service.

Questions

10. Are there any additional administrative costs that are charged separately and are not covered in this RFP (for example, special management reports, billing data, and any other special services)?

Glossary

Abstract Syntax Notation (ASN.1) A coding scheme used in international standards such as X.400 and X.500.

Access unit (AU) A component of an X.400 network that enables users of one service to communicate with message handling services, such as the Interpersonal Messaging (IPM) service.

Actual recipient In X.400, the recipient for which delivery or notification of receipt takes place.

Address A unique identifier of a mailbox.

Administration In X.400, a commercial administration or a recognized private operating agency (RPOA).

Administration domain name In X.400, the standard attribute that identifies the country in which an ADMD is located.

Administrative Directory Management Domains (ADDMDs) ADDMDs are part of the X.500 directory, managed by an organization such as a government agency, an educational institution, or a business.

Administrative Domain (ADMD) In X.400, an organization that offers messaging transport services to individuals or organizations. MCI and AT&T are examples of ADMDs in the United States.

Alternate recipient A user or distribution list to which the originator wants the message to be conveyed if—and only if—the system cannot deliver it to the preferred recipient.

American National Standards Institute (ANSI) An organization that sets standards in the United States. ANSI also registers domain names.

American Standard Code for Information Interchange (ASCII) The character coding system most commonly used by computer systems.

Apple Open Collaborative Environment (AOCE) A set of technologies available from Apple Computer that is designed to consolidate workgroups and workflow within a network environment.

Application Program Interface (API) Software toolkits that enable a programmer to build links to a specific application. The three most prevalent APIs in e-mail are Common Message Calls (CMC), Message Application Program Interface (MAPI), and Vendor Independent Messaging (VIM).

Attachment   Output of a computer application program, such as a word processing document, that is appended to a message.

Attribute   A component of an attribute list that describes a user or distribution list.

Attribute list   A data structure that constitutes an originator/recipient (O/R) address.

Attribute type   A part of an attribute that identifies a class of information, such as personal names.

Attribute value   Information supplied to complete the attribute type. For example, an attribute type of "country" could have an attribute value of "US" or "GB."

Auto-answer   A mechanism on the user's message store that automatically responds to received messages with a short predetermined message.

Auto-forward   A mechanism on the user's message store that automatically re-sends a received message to a predetermined address.

Automated Directory Update (ADE)   The directory update protocol used in cc:Mail.

Basic service   In the X.400 standard, the sum of the features offered by a messaging service.

Body   One of the four possible components of a message. The other components are the heading, attachment, and envelope.

Body part   One of the parts of the body of a message.

Character set   A compilation of alphanumeric characters that may be used in the parts of a message. The address may allow the use of character sets different than those used in the message body. Examples are ASCII and IA5.

Common Messaging Calls (CMC)   An API produced by an industry consortium called X.400 API Association (XAPIA).

Common name   In X.400, a standard part of an originator/recipient (O/R) address form that identifies a user or distribution list, such as "Ben Smith" or "All Managers." For example, the components of a common name might include a given name, initials, a surname, and a generation qualifier.

Content   The subject matter contained within the message. In an SMTP message, content is equivalent to an X.400 body. An X.400 body may have several parts, each of which may be called a content. In X.400, the content is neither examined nor modified, except for conversion, during the conveyance of the message.

Content type   In X.400, an identifier on a message envelope that states the type of information in the content, such as Binary or Group 3 fax.

Conversion   An event in which a message transfer agent (MTA) transforms parts of a message's content or a body part from one encoded information type to another, such as from spreadsheet type "Excel" to spreadsheet type "Lotus123."

Country name   A standard attribute of an X.400 and an X.500 name form that identifies a country. A country name is a unique designation used for the purpose of sending and receiving messages.

Delivery   A message transfer step in which a message transfer agent (MTA) conveys a message to the message store (MS), message user agent (MUA), or access unit (AU) of a potential recipient.

Delivery report   A report that acknowledges delivery, nondelivery, export, affirmation of the subject message or probe, or distribution list expansion.

Directory   A collection of systems cooperating to provide directory services.

Directory Access Protocol (DAP)   An X.500 protocol used by the directory user agent (DUA) to access the directory system agent (DSA).

Directory Information Base (DIB)   A collection of entries describing single objects in an X.500 directory.

Directory Information Shadowing Protocol (DISP)   The rules and syntax that define how X.500 directories are replicated across distributed, cooperating directory system agents (DSAs).

Directory Information Tree (DIT)   A hierarchical model used to organize X.500 directory information, such as a user's e-mail address.

Directory system agent (DSA)   A part of a directory whose function is to enable access by DUAs and/or other DSAs to the directory information base.

Directory system agent (DSA)   A database containing X.500 data. The X.500 directory comprises many such DSAs.

Directory System Protocol   An X.500 protocol used by a directory system agent (DSA) to exchange information and requests with other DSAs.

Directory user agent (DUA)   An application process that represents a user in accessing the directory. Each DUA serves a single user so that the directory can control access to directory information on the basis of the DUA names.

Distribution group name/distribution element name (DGN/DEN)   The SNADS addressing string.

Distribution list (DL)   An object that represents a prespecified group of users and other distribution lists that may be addressed by name.

Distribution list expansion   An event in which an MTA expands a distribution list into its component addresses.

Distribution list name   The name allocated to a distribution list.

Document Interchange Architecture (DIA)   An IBM architecture that establishes the structure and protocols required to exchange program data over a network.

Document interchange unit (DIU)   Part of the IBM Document Interchange Architecture, composed of a message body, a header, and an envelope designed and structured for transmission over a network.

Domain-defined attribute   In X.400, an optional, special part of the originator/recipient (O/R) name that is used by the destination delivery system.

Electronic Data Interchage (EDI)   The paperless exchange of business documents, such as invoices, from a computer in one organization to a computer in another.

Electronic Messaging Association (EMA)   An association of vendors, government entities, corporations, and industry trade groups that are active in the messaging industry.

Element of service   A describable part of all the services offered by an e-mail system.

Encoded information type (EIT)   In X.400, an identifier on a message envelope that indicates one type of information contained in the message body, such as IA5 text or Group 3 fax.

Envelope   A component of a message used by MTAs to convey and handle the message. In many, but not all, systems, the envelope is the only part of the message that is modified by the MTAs. The other message components are the heading, attachment, and body.

Explicit conversion   A type of conversion in which the originator selects both the initial and final encoded information types.

Gateway   A software application program that enables disparate e-mail systems to transparently exchange messages, files, and attachments to the end user.

Global Message Handling Service (Global MHS)   A message transport system used in Novell's LAN environment.

Heading   A component of an electronic message that contains addressing information. The other components are the envelope, attachment, and body.

Implicit conversion   In X.400, a type of conversion in which the message transfer agent (MTA), not the user, selects both the initial and final encoded information types.

Indirect submission   In X.400, a transmittal step in which an originator's message user agent (MUA) gives the message to the message store (MS) for submission to the message transfer agent (MTA).

International Alphabet Number 5 (IA5)   A type of text, which is defined in the international standard recommendation T.50.

International Organization for Standardization (ISO)   An organization that is responsible for developing standards for the international exchange of almost all services and manufactured products.

International Telecommunications Union (ITU)   A United Nations organization responsible for a committee, called the Consultative Committee for International Telegraphy and Telephony (CCITT), that recommends telecommunications standards.

Interpersonal Messaging (IPM)   E-mail between two or more end users, as defined in the X.400 messaging recommendations.

Local area network (LAN)   Hardware and software used to connect terminals, workstations, servers, and hosts into a single network environment.

Mailbox storage   An e-mail network function for storing inbound user messages.

Management domain (MD)   In the context of message handling, a set of messaging systems—at least one of which contains, or recognizes, an MTA—that is managed by a single organization. The MD is a primary building block used in the organizational construction of an MHS and refers to an organizational area for the provision of services.

Management domain name   The unique designation of a management domain for the purpose of sending and receiving messages.

Masquerade   A type of security threat that occurs when an entity successfully pretends to be a different one.

Message   An electronic communication transmitted from one end user to one or more recipients. It is usually composed of an envelope, heading, attachment, and body.

Message Handling Service (MHS)   A Novell product for the exchange of messages in a local area network environment.

Message Handling System (MHS)   All the parts that compose an electronic messaging environment: the message store (MS), message user agent (MUA), and message transfer agents (MTAs).

Message store (MS)   A component of a message handling system (MHS) that provides a single direct user with capabilities for message storage. It is always available to message transfer agents (MTAs) for the delivery of messages.

Message switch   An integrated multiple-message gateway.

Message transfer agent (MTA)   A component of a message transfer system (MTS) that actually conveys information objects to users and distribution lists.

Message transfer system (MTS)   A system consisting of one or more message transfer agents (MTAs) that provide store-and-forward message transfer among message user agents (MUAs), message stores (MSs), and access units (AUs).

Message user agent (MUA)   Often simply referred to as the user agent (UA), the MUA is the interface between the end user of an e-mail system and the message transport agent (MTA).

Messaging Application Program Interface (MAPI)   A toolkit available from Microsoft that enables developers to generate links between Microsoft Mail and various application programs.

Modification   A type of security threat that occurs when a message is altered in an unanticipated manner by an unauthorized entity.

Multipurpose Internet Mail Extension (MIME)   An enhancement to SMTP that permits Internet e-mail to handle multiple body parts.

Naming authority   An entity responsible for the allocation of names within an organization.

National Institute of Standards and Technologies (NIST)   Formerly the National Bureau of Standards (NBS), NIST controls the Federal Information Processing Standards (FIPS), selects standards for the government, and defines profiles for open systems interconnection.

Network address   In X.400, a standard attribute of an O/R address form that provides a terminal's network address.

Nondelivery   An event in which an MTA determines that it cannot deliver a message to one or more of its immediate recipients.

Office System Node (OSN)   The end point of a system; used in PROFS and Personal Services/SL.

Open Systems Interconnection (OSI) model   The International Organization for Standardization's (ISO) seven-layer model for data communications between compliant computer systems. The model is comprised of seven distinct functions, such as network and application functions.

Organization name   A standard attribute of an originator/receipient (O/R) address that uniquely identifies an organization for the purpose of sending and receiving messages.

Organizational unit name   A standard attribute of an originator/receipient (O/R) address that uniquely identifies an organizational unit of an organization for the purpose of sending and receiving of messages.

Originator   The user who is the ultimate source of a message.

Originator/recipient (O/R) address   The originator/recipient name format, called Variant 1.1, used most often in X.400, which is an attribute that distinguishes one user or distribution list from another.

Originator/recipient (O/R) name   An originator/recipient name that distinguishes one user or distribution list from another.

OSI Directory   The name of an X.500 directory.

Personal name   In X.400, a standard attribute of an originator/receipient (O/R) address form that identifies a person.

Post office (PO)   A component of a message handling system (MHS), sometimes called a message store (MS), that temporarily holds inbound and outbound mail and usually resides on the server. It is always available to message transfer agents (MTAs) for the delivery of messages.

Private Directory Management Domain (PRDMD)   In X.500, the entity responsible for administering the directory system.

Private domain name   In X.400, a standard attribute of an originator/receipient (O/R) address form that identifies a PRMD relative to the ADMD denoted by an administration domain name.

Private Management Domain (PRMD)   In X.400, a management domain that comprises a messaging system (or systems) managed and operated by an organization other than an administration or RPOA.

Protocol 1 (P1)   A protocol defined in the X.400 standard that is used between MTAs.

Protocol 2 (P2)   A protocol defined in the X.400 standard that is used between e-mail message user agents (MUAs).

Protocol 3 (P3)   A protocol defined in the X.400 standard that is used between the message store (MS) or the message user agent (MUA) and the message transport agent (MTA).

Protocol 7 (P7)   A protocol defined in the X.400 standards that is used between the message user agent (MUA) and the message store (MS).

Receipt   A message transmittal step in which either a message user agent (MUA) conveys a message or report to its direct user, or the communication system that serves as an indirect user conveys such information to that user.

Relative Distinguished Name (RDN)   A unique set of attributes and values for an entry in an X.500 directory.

Replay   A type of security threat that occurs when an exchange is captured and re-sent at a later time to confuse the original recipients.

Report   An information object generated by the MTS that conveys the outcome or progress of a message's transmittal to one or more potential recipients.

Request For Comment (RFC)   A document used in the Internet community to request that users comment on proposed standards and other activities.

Security capabilities   In the context of message handling, the mechanisms that protect against various security threats.

Simple Mail Transfer Protocol (SMTP)   The protocol used on the Internet network to move messages between systems.

Standard Message Format, Version 71 (SMF71)   The newest Novell message structure definition.

Subject   The information in a header that summarizes the contents of a message, as specified by the originator.

Systems Network Architecture Distribution Services (SNADS)   This service provides the transfer of messages and files among users in an IBM SNA network.

Traffic analysis   A type of security threat that occurs when an outside entity is able to monitor and analyze traffic patterns on a network.

Transfer   In the context of message handling, a transmittal step in which one MTA conveys a message, probe, or report to another.

Transmission Control Protocol/Internet Protocol (TCP/IP)   The Internet equivalent of the OSI model's Network layer 3 and Transport layer 4.

User   A component of the message handling environment, that engages in (rather than provides) message handling and is a potential source or destination for the information objects conveyed by a message handling system (MHS).

User agent (UA)   Often referred to as a message user agent (MUA), the component of a message handling system (MHS) with which the user interacts to perform messaging tasks.

Value added network (VAN)   A commercial electronic messaging network that moves e-mail or EDI messages among subscribers for a fee.

Vendor Independent Messaging (VIM)   An API developed by a consortium led by Lotus Development Corporation, used to interface cc:Mail to various applications.

Voice body part   A body part sent or forwarded from an originator to a recipient that conveys voice-encoded data and related information. The related information consists of parameters that help process the voice data, such as its duration and the algorithm used to encode it.

X.400   A set of recommendations defined by the International Telecommunications Union that describe the X.400 Message Handling System.

X.400 API Association (XAPIA)   A standards group that is actively defining standard application program interfaces (APIs) for X.400 messaging systems.

X.435   A set of recommendations defined by the International Telecommunications Union that describe the Message Handling System to support EDI.

X.500   A set of recommendations defined by the International Telecommunications Union that describe a globally distributed electronic directory system.

Zone Information Protocol 5 (ZIP5)   The ZIP5 message format is used by PROFS systems to send e-mail. (PROFS message packet is a sequential group of 80-character punch card-like strings.)

Bibliography

Administrators' Guide, DaVinci eMail Version 2.5 for DOS and Windows. DaVinci Systems, Inc., 1993.

Alvestrand, H., S. Kille, R. Miles, M. Rose, and S. Thompson. *Mapping between X.400 and RFC 822 Message Bodies,* RFC 1495.

Alvestrand, H., and S. Thompson. *Equivalencies between 1988 X.400 and RFC 822 Message Bodies,* RFC 1494.

Anderson, M. "Electronic Mail Cost of Ownership." *GartnerGroup OSI Research Notes,* August 9, 1993.

Anderson, M. "OIS Migration: Cost Analysis Results." *GartnerGroup Research Notes,* August 9, 1993.

Betanov, Cemil. *Introduction to X.400.* Boston: Artech House, 1993.

Borenstein, N., and N. Freed. *MIME (Multipurpose Internet Mail Extentions) Part One: Mechanisms for Specifying and Describing the Format of Internet Message Bodies,* RFC 1521.

cc:Mail Automatic Directory Exchange Reference Manual. Mountain View, Calif.: Lotus Development Corporation, 1991.

cc:Mail Directory Services User's Manual, Version 1.0. Mountain View, Calif.: Lotus Development Corporation, 1991.

The cc:Mail E-mail Comparison Guide. Mountain View, Calif.: Lotus Development Corporation, 1993.

cc:Mail for MS-DOS Administrator's Manual. Mountain View, Calif.: Lotus Development Corporation, 1992.

cc:Mail Import/Export User's Manual, Version 3.3. Mountain View, Calif.: Lotus Development Corporation, 1992.

cc:Mail Router Administrator Guide, Version 4.0. Mountain View, Calif.: Lotus Development Corporation, 1993.

Common Messaging Call API, Version 1.0. Mountain View, Calif.: X.400 API Association, 1993.

Course 1770, Administering SMTP for NetWare Global MHS Self-Paced Manual. Provo, Utah: Novell, Inc., 1993.

Crocker, David H. *Standard for the Format of ARPA Internet Text Messages,* RFC 822.

Drummond, Richard V., ed. *Data Communications for the Office.* New York: Bantam Professional Books, 1993.

Drummond, R. *Request for Proposal E-mail Switches.* Drummond Group, February 1994.

Emc^2/*TAO Catalog of Features, Version 3.3.* Fischer International Systems Corporation, 1993.

Ferris, David. "The Business Justification of E-mail." *Ferris E-mail Analyzer* 2, no. 8, August 1993.

Getting Started with OfficeVision/MVS, Release 2.0. Roanoke, Tex.: International Business Machines Corporation, 1990.

Gillooly, Caryn. "User Firms Plagued by LAN Money Pits." *Network World,* January 4, 1993, 1.

"Grading E-mail." *Network World,* September 21, 1992, 39.

Hardcastle-Kille, S. *Mapping Between X.400 (88)/ISO 10021 and RFC 822,* RFC 1327.

IBM Application System/400 Communications: Distributed Services Network Guide, Version 2. Rochester, N.Y.: International Business Machines Corporation, 1992.

IBM Application System/400 Office Services Concepts and Programmer's Guide, Version 2. Roanoke, Tex.: International Business Machines Corporation, 1993.

IBM Document Interchange Architecture Technical Reference. Roanoke, Tex.: International Business Machines Corporation, 1989.

IBM Systems Network Architecture Distribution Services Reference. Research Triangle Park, N.C.: International Business Machines Corporation, 1989.

IBM Using OfficeVision/VM Release 2.0. Roanoke, Tex.: International Business Machines Corporation, 1992.

Intelligent Messaging Mail Administrator's Guide. Banyan Systems Inc., 1992.

International Telecommunications Union, CCITT, Data Communications Networks: Open Systems Interconnection (OSI); Security, Structure, and Applications, X.800. Geneva: IXth Plenary Assembly, 1991.

International Telecommunications Union, CCITT, Volume VIII - Fascicle VIII.7, Data Communication Networks Message Handling Systems, Recommendation, X.400-X.420. Geneva: IXth Plenary Assembly, 1989.

International Telecommunications Union, CCITT, Volume VIII - Fascicle VIII.8, Data Communication Networks Directory, Recommendation, X.500-X.521. Geneva: IXth Plenary Assembly, 1988.

International Telecommunications Union, CCITT, Volume VIII - Fascicle VIII.8, Data Communication Networks Directory, Recommendation, X.509. Geneva: IXth Plenary Assembly, 1988.

Introduction to MAILbus. Maynard, Massachusetts: Digital Equipment Corporation, 1989.

Keypak ODX Conversion Reference Guide, Version 3.1. Keyword Office Technologies, 1994.

Kille, S., and N. Freed. *Mail Monitoring MIB*, RFC 1566.

Klensin, J., N. Freed, and K. Moore. *SMTP Service Extensions for Message Size Declaration*, RFC 1427.

Klensin, J., N. Freed, M. Rose, E. Stefferud, and D. Crocker. *SMTP Service Extensions*, RFC 1425.

Klensin, J., N. Freed, M. Rose, E. Stefferud, and D. Crocker. *SMTP Service Extensions for 8bit-MIME Transport*, RFC 1426.

Lotus cc:Mail Third Party Catalog. Cambridge, Mass.: Lotus Development Corporation, 1993.

Lotus Message Strategy, New Communications Product Direction White Paper. Cambridge, Mass.: Lotus Development Corporation, 1993.

Lotus Notes Administrator's Guide for OS/2 and UNIX, Release 3. Cambridge, Mass.: Lotus Development Corporation, 1993.

Lotus Notes Administrator's Guide, Lotus Notes Server for Windows Release 3. Cambridge, Mass.: Lotus Development Corporation, 1993.

Lotus Notes Administrator's Guide, Release 3. Cambridge, Mass.: Lotus Development Corporation, 1993.

Lotus Notes Application Developer's Reference, Release 3. Cambridge, Mass.: Lotus Development Corporation, 1993.

Lotus Notes Concepts Release 3 Student Handbook. Cambridge, Mass.: Lotus Development Corporation, 1993.

Lotus Notes Release 3, Application Developer's Reference. Mountain View, Calif.: Lotus Development Corporation, 1993.

Lotus Notes Release 3, Getting Started with Application Development. Cambridge, Mass.: Lotus Development Corporation, 1993.

Lotus Notes Technical User Release Student Handbook. Mountain View, Calif.: Lotus Development Corporation, 1993.

Lotus VIM Developer's Toolkit for cc:Mail and Lotus Notes. Cambridge, Mass.: Lotus Development Corporation, 1993.

Lotus VIM Developer's Toolkit for cc:Mail and Lotus Notes Programmer's Guide. Mountain View, Calif.: Lotus Development Corporation, 1993.

McCoy, E., and R. Freiwirth. *Electronic Mail Management Requirements, Joint IFIP WG6.5 and 6.6 Electronic Mail Management Working Group,* October 1993.

McGarr, Mike. "1994 VAN/VAB Survey." *EDI World,* January 1994.

MHS Directory Administrator's Guide Advanced Release, Version 5.8. General Electric Company, 1991.

Microsoft Mail Administrator's Guide for PC Networks, Version 3.2. Redmond, Wash.: Microsoft Corporation, 1993.

Microsoft Mail Technical Reference for PC Networks, Version 3.2. Redmond, Wash.: Microsoft Corporation, 1993.

Moore, K. *MIME (Multipurpose Internet Mail Extensions) Part Two: Message Header Extensions for Non-ASCII Text,* RFC 1522.

NetWare Global Messaging Administration. San Jose: Novell, Inc., 1992.

Notes Application Development II Student Handbook, Release 3. Cambridge, Mass.: Lotus Development Corporation, 1993.

Notes Release 3 Update Student Handbook. Mountain View, Calif.: Lotus Development Corporation, 1993.

Notes System Administrator 1, Release 3, Student Handbook. Cambridge, Mass.: Lotus CSG Education Services, 1993.

OpenMail Technical Guide. Hewlett Packard, 1993.

Postel, Jonathan B. *Domain Name System Structure and Delegation,* RFC 1591.

Postel, Jonathan B. *Simple Mail Transfer Protocol,* RFC 821.

QuickMail Administrator Manual. CE Software, Inc., 1993.

QuickMail Getting Started with QuickMail. CE Software, Inc., 1993.

QuickMail QM Forms User Manual. CE Software, Inc., 1990.

QuickMail User Guide for DOS. CE Software, Inc., 1991.

QuickMail User Manual for Macintosh. CE Software, Inc., 1991.

QuickMail User Manual for Windows. CE Software, Inc., 1993.

Rose, Marshall T., and Dwight E. Cass. *ISO Transport Services on Top of the TCP, Version 3,* RFC 1006.

Rose, Marshall T. *The Open Book, A Practical Perspective on OSI.* Englewood Cliffs, N.J.: Prentice Hall, 1990.

Software Developer's Kit SMF v71 Programmer's Reference. San Jose: Novell, Inc., 1992.

Systems Application Architecture OfficeVision/400, Using OfficeVision/400 Version 2. Roanoke, Tex.: International Business Machines Corporations, 1993.

Teamlinks Handbook. Digital Equipment Corporation, 1993.

Technical Reference, Microsoft Mail Electronic Mail for PC Networks Version 3.2. Redmond, Wash.: Microsoft Corporation, 1993.

Treacy, Michael E. *The Cost of Network Ownership Index Group,* 1989.

User's Guide, DaVinci eMail Version 2.0 for DOS. DaVinci Systems, Inc., 1991.

User's Guide, Microsoft Mail Electronic Mail for PC Networks Workstation Software for MS-DOS Version 3.0. Redmond, Wash.: Microsoft Corporation, 1992.

User's Guide to Microsoft Mail Electronic Mail for PC Networks Workstation Software for Windows and Presentation Manager Version 3.0. Redmond, Wash.: Microsoft Corporation, 1993.

VAX Message Router Programmer's Kit Progammer's Guide, Version 3.3. Digital Equipment Corporation, 1993.

VAX Message Router VMSMail Gateway User's Guide, Version 3.2. Maynard, Mass.: Digital Equipment Corporation, 1991.

VINES Intelligent Messaging Mail User's Guide. Banyan Systems, Inc., 1992.

VINES Mail Client Progamming Interface. Banyan Systems, Inc., 1993.

VINES Mail Gateway Programming Interface. Banyan Systems, Inc., 1993.

Wegmann, C. "Buy a LAN and Pay for Three." *GartnerGroup Research Notes,* August 9, 1993.

Index

Symbols

; (semicolon), in X.400 addressing, 228
/ (slash), in X.400 addressing, 228

A

Abstract Syntax Notation (ASN.1) encoding, 233-236, 239-240
Access control, 276-277
Access units, 24-25
Accounting management, 252
Act On CMC function, 287, 289
ADDMD (Administrative Directory Management Domain), X.500, 180
Address books, 24, 143-144
Address definition call, CMC, 288
Address directories
    Simple Message Transfer Protocol (SMTP), 206, 217
    SNA Distribution Services (SNADS), 192
    X.400 standard, 241
Address mapping, 159-160, 164
    See also Mapping
Addresses
    defined, 5
    Vendor Independent Messaging (VIM) API and, 286

    See also Contact information
Addressing
    across VANs, 58-59, 226-228
        Internet addressing, 59
        proprietary addressing, 59
        X.400 addressing, 59, 226-228
    in ALL-IN-1, 89-90, 93, 95
    in NetWare Global Message Handling Service (Global MHS), 194-195
    in SNADS and DIA, 188
    in X.400 standard, 59, 222-228
        Form 1, Variant 1 addressing, 223-226
        between systems, 59, 226-228
        types of, 222-223
Addressing structures
    ALL-IN-1, 89-90, 93, 95
    BeyondMail, 68
    cc:Mail, 135-138
    DaVinci eMail, 85-86
    Emc^2/TAO, 107
    Lotus Notes, 144-145
    Microsoft Mail, 152, 155
    OfficeVision/400, 129
    OfficeVision/VM, 123
    Open DeskManager, 115-116
    QuickMail, 77-78
    Simple Message Transfer Protocol (SMTP), 206-207

Administration calls, CMC, 287
Administration tools, 48
Administrative Directory Management Domain
    (ADDMD), X.500, 180
Administrative specifications, in RFP/RFI
    example, 310-311
ALL-IN-1 software, 87-103
        addressing, 89-90, 93
        addressing structure, 95
        attachments, 92
        compound document support, 91
        contact information, 87
        delivery notification, 93
        directory structure, 95
        Distributed Directory Services (DDS), 95
        file cabinet feature, 89, 93, 102
        foreign language support, 90
        gateways supported, 102-103
        Group Conferencing feature, 91
        message functions, 92
        Message Router OpenVMS gateway, 90,
            95-97
        message store, 102
        message structure, 97-102
            body part types, 101-102
            heading fields, 100-101
            name format, 98-100
            optional elements, 98
            overview of, 97-98
        message transport system, 95-97
        message user agent, 94
        network specifications, 94
        online training feature, 91
        operating system specifications, 93
        overview of, 87-89
        platform specifications, 93
        printing features, 90, 91
        receipt notification, 93
        Set Mail User option, 89
        shared filing, 89
        standards support, 93
        System Management utilities, 91
        TeamLinks support, 91-92
        Time Management feature, 90
        user profile feature, 90
Alternate recipient allowed services, 166
Annotations, document conversion and, 272
Application layer, OSI protocol stack, 245-246
Application program interfaces (APIs), 282-292
        Common Messaging Calls (CMC), 286-290
            Act On function, 289
            address definition call, 288
            administration calls, 287
            List function, 289
            looking up names, 287
            message definition call, 288
            messaging calls, 287-289

        overview of, 286-287
        Read function, 289
        receiving messages, 287
        Send function, 288-289
        sending messages, 287
    Messaging Application Program Interface
        (MAPI), 290-292
        MAPIAddress function, 291
        MapiFileDesc definition, 291-292
        MapiMessage structure definition, 291, 292
        MapiRecipDesc definition, 291
        MAPISendMail function, 291
        overview of, 290-291
    overview of, 282
    Vendor Independent Messaging (VIM),
        282-286
        address format, 286
        character sets, 283
        data attribute encoding, 283
        message creation process, 284-286
        message structure, 283
        overview of, 282-283
        Simple Message Interface (SMI) APIs,
            283-284
        VIMCreateMessage function, 284
        VIMSetMessageHeader function, 284-285
        VIMSetMessageItem function, 285-286
        VIMSetMessageRecipient function, 285
ASCII data, converting binary data to, 211, 215-217
ASN.1 (Abstract Syntax Notation) encoding,
    233-236, 239-240
Attachments. *See* Message attachments
Authenticated and authorized users, 276
Autoforward indication services, 166

B

Backbones, 260
Backup specifications, in RFP/RFI example, 308
Banyan Systems, Inc. *See* BeyondMail software
BeyondMail software, 63-74
        addressing structure, 68
        BeyondRules scripting language, 65
        body structure fields, 72-73
        contact information, 63
        DDE support, 65
        directory structure, 66-67
        directory synchronization services, 67
        editor, 64
        envelope fields, 70-72
        gateways supported, 73-74
        intelligent features, 65
        message store, 73
        message structure, 69-73
        message transport system, 68-69, 70
        network specifications, 65-66

OLE support, 65
operating system specifications, 65-66
Outside/In Views, 64
overview of, 63-64
platform specifications, 65-66
security, 64
user agent, 66
Windows features, 65
Binary data, converting to ASCII data, 211, 215-217
Blind copy recipient handling services, 167
Body. *See* Message body
Bulletin board support
Emc^2/TAO, 105
Microsoft Mail, 149
OfficeVision/400, 128
OfficeVision/VM, 121

C

Calendaring features
ALL-IN-1, 90
cc:Mail, 133
Emc^2/TAO, 104
Microsoft Mail, 149
OfficeVision/400, 127-128
OfficeVision/VM, 120
CallUp directory, OfficeVision/VM, 122
Carriers. *See* Value added networks
cc:Mail software, 130-140
addressing structure, 135-138
body part types, 140
contact information, 131
directory structure, 134-135
directory updating mechanism, 137-138
gateways supported, 140
heading fields, 139
for Macintosh, 133
message store, 140
message structure, 139-140
message transport system, 138-139
message user agent, 134
mobile mail support, 133
for MS-DOS, 132
network specifications, 134
operating system specifications, 133
for OS/2 Workplace Shell, 132-133
overview of, 130-133
platform specifications, 133
time management feature, 133
for UNIX, 133
for Windows, 132
CE Software. *See* QuickMail software
Chaining, in X.500 directory standard, 183-184
Character information, document conversion and, 272
Character set mapping, 164

Character sets, VIM APIs and, 283
CMC API. *See* Common Messaging Calls (CMC) API
Columns, document conversion and, 272
Common Messaging Calls (CMC) API, 286-290
Act On function, 287, 289
address definition call, 288
administration calls, 287
List function, 287, 289
looking up names, 287
message definition call, 288
messaging calls, 287-289
overview of, 286-287
Read function, 287, 289
receiving messages, 287
Send function, 287, 288-289
See also Application program interfaces
Communications carriers. *See* Value added networks (VANs)
Compound documents, in ALL-IN-1, 91
Conferencing features
ALL-IN-1, 91
Emc^2/TAO, 105
Configuration management, 252
Connecting e-mail systems. *See* Interoperability
Contact information
ALL-IN-1, 87
BeyondMail, 63
cc:Mail, 131
DaVinci eMail, 82
Emc^2/TAO, 103
Lotus Notes, 141
Microsoft Mail, 149
OfficeVision/400, 126
OfficeVision/VM, 119
Open DeskManager, 109
QuickMail, 74
Contents
content information services, 166
content type services, 167
return of content with nondelivery notification services, 169
subject field content services, 165
Content-Transfer-Encoding header field, SMTP, 215-217
Conversion prohibition services, 167-168
Converted indication services, 168
Converting
binary data to ASCII data, 211, 215-217
documents through gateways, 49
Cooperative queries, 183-184, 185
Costs, 32, 33-38
Ferris Network LAN e-mail cost study, 33-35
Forrester LAN cost study, 33
GartnerGroup e-mail cost study, 35-36
GartnerGroup LAN cost study, 36
GartnerGroup OIS migration cost study, 36

in RFP/RFI example, 312-315
Cross-referencing indication services, 168
Customer support specifications, in RFP/RFI
example, 311-312

D

DAP (Directory Access Protocol), 179
Data attribute encoding, VIM APIs and, 283
Data-link layer, OSI protocol stack, 247-248
DaVinci eMail, 82-87
addressing structure, 85-86
contact information, 82
directory structure, 85
foreign language support, 83
gateways supported, 87
message store, 86
message structure, 86
message transport system, 86
network specifications, 85
operating system specifications, 85
overview of, 83-84
platform specifications, 85
related products, 84-85
user agent, 85
DDE support, in BeyondMail, 65
DDS (Distributed Directory Services), ALL-IN-1, 95
DEC. *See* ALL-IN-1 software
Delivery. *See* Message delivery
DIA. *See* Document Interchange Architecture (DIA)
Digital Equipment Corporation. *See* ALL-IN-1
software
Directories
address directories
Simple Message Transfer Protocol (SMTP),
206, 217
SNA Distribution Services (SNADS), 192
X.400 standard, 241
centralized directory facilities on gateways,
45
defined, 5
directory and transport backbone gateway
components, 260
directory synchronization with gateways, 45
shared directories, 16-19
global distributed directories, 18
global nondistributed directories, 18
maintaining global directory information,
18-19
overview of, 16-17
See also Directory structures
Directory Access Protocol (DAP), 179
Directory Information Shadowing Protocol (DISP),
179
Directory information tree (DIT), X.500 directory
standard, 181

Directory queries, 183
Directory structures
ALL-IN-1, 95
BeyondMail, 66-67
cc:Mail, 134-135
Emc^2/TAO, 106-107
Lotus Notes, 143-144
Microsoft Mail, 151-152
NetWare Global Message Handling Service
(Global MHS), 201-203
OfficeVision/400, 128-129
OfficeVision/VM, 122-123
Open DeskManager, 111-115
QuickMail, 76-77, 78-80
See also Directories
Directory synchronization services
BeyondMail, 67
OfficeVision/VM, 122
Directory system agents (DSAs), 18, 183-184, 185
Directory System Protocol (DSP), 179
Directory updating
cc:Mail, 137-138
Microsoft Mail, 152
Directory user agents (DUAs), 19, 183-184, 185
DISP (Directory Information Shadowing Protocol),
179
DisplayWrite/370 program, OfficeVision/VM,
121-122
Distinguished names, in X.500 directory standard,
182
Distributed Directory Services (DDS), ALL-IN-1, 95
Distributed multisystem gateways, 44, 46-47
Distributed Services (DS) protocol, Open
DeskManager, 116
Distribution manager, OfficeVision/VM, 124
Distribution service units (DSUs), 21
DIT (directory information tree), X.500 directory
standard, 181
DL expansion history indication, 169-170
DL expansion prohibited services, 170
DNS (Domain Name Service) addressing
structure, SMTP, 206-207
Document conversion, 268-273
columns and, 272
conversion features, 271-273
document information and, 271
fields and, 273
frames and, 272
gateways and, 49
graphics and, 273
information loss and, 270
KEYPak software, 270-271
in Microsoft Mail, 150
multiple columns and, 272
notes and, 272
in OfficeVision/VM, 121-122
overview of, 268-269

page layout and, 271
   paragraph information and, 272
   section information and, 271
   tables and, 272
Document Interchange Architecture (DIA), 188-192
   addressing in, 188
   message structure, 189-192
   message transport system, 189
   See also SNA Distribution Services
Documents
   compound documents in ALL-IN-1, 91
   in OfficeVision/400, 127
   in OfficeVision/VM, 120-121
   See also Messages; Notes
Domains
   SMTP Domain Name Service (DNS), 206-207
   top level domains (TLDs), 206-207
   X.500 Administrative Directory
      Management Domain (ADDMD), 180
   X.500 Private Directory Management
      Domain (PRDMD), 180
DOS, cc:Mail version for, 132
DOS Office Direct Connect feature,
   OfficeVision/VM, 121
DS (Distributed Services) protocol, Open
   DeskManager, 116
DSAs (directory system agents), 18, 183-184, 185
DSP (Directory System Protocol), 179
DSUs (distribution service units), 21
DUAs (directory user agents), 19, 183-184, 185

E

EBPs (extended body parts), 162
Editor, in BeyondMail, 64
Electronic Data Interchange (EDI), VANs and, 54,
   57
E-mail
   costs, 32, 33-38
      Ferris Network LAN e-mail cost study,
         33-35
      Forrester LAN cost study, 33
      GartnerGroup e-mail cost study, 35-36
      GartnerGroup LAN cost study, 36
      GartnerGroup OIS migration cost study, 36
      other studies, 37-38
      in RFP/RFI example, 312-315
   defined, 5
   organizational impact of, 32
   overview of, 4-7
E-mail connectivity. *See* Interoperability
E-mail delivery. *See* Message delivery
E-mail gateways. *See* Gateways
E-mail integration. *See* Interoperability
E-mail network management. *See* Management;
   Network management

E-mail switches, 6-7, 43-44
E-mail systems, 4-7, 14-30
   access units, 24-25
   directory system agents (DSAs), 18, 183-184,
      185
   directory user agents (DUAs), 19, 183-184,
      185
   distribution service units (DSUs), 21
   mainframe-based e-mail systems, 36
   message transport agents (MTAs), 14-16,
      20-21, 222
   overview of, 4-7, 14-16
   personal address books, 24, 143-144
   shared directories, 16-19
      global distributed directories, 18
      global nondistributed directories, 18
      maintaining global directory information,
         18-19
      overview of, 16-17
   See also entries for specific software packages;
      Gateways; Interoperability; Message
      store; Message structures; Message
      transport systems (MTSs); Message user
      agents (MUAs)
Emc²/TAO software, 103-108
   addressing structure, 107
   calendaring features, 104
   conferences and bulletin board support, 105
   contact information, 103
   directory structure, 106-107
   foreign language support, 105
   forms feature, 105
   gateways supported, 108
   message structure, 108
   message transport system, 108
   message user agent, 106
   network specifications, 105
   operating system specifications, 105
   overview of, 103-104
   platform specifications, 105
EMS (express mail service), 170
Encryption, 48, 167, 276-277
End-to-end security, 276, 278
Envelopes. *See* Message envelopes
Express mail service (EMS), 170
Extended body parts (EBPs), 162
Extended Simple Message Transfer Protocol
   (SMTP), 210-211
EXTERNAL program, Microsoft Mail, 153-154

F

Fault management, 252
Fax, testing delivery via, 169
Ferris Network LAN e-mail cost study, 33-35
Fields, document conversion and, 273

File cabinet feature
    ALL-IN-1, 89, 93, 102
    Microsoft Mail, 150
    OfficeVision/VM, 121
    Open DeskManager, 111
File formats, in OfficeVision/VM, 123
Fischer International. *See* Emc^2/TAO software
Foreign language support
    ALL-IN-1, 90
    DaVinci eMail, 83
    Emc^2/TAO, 105
Form 1, Variant 1 addressing, X.400, 223-226
Forms features
    Emc^2/TAO, 105
    Microsoft Mail, 149-150
    Open DeskManager, 111
Forrester LAN cost study, 33
Forward slash (/), in X.400 addressing, 228
Frames, document conversion and, 272
Frameworks, for network management, 251-252
Function calls. *See* Application program interfaces
    (APIs)
Functionality tests. *See* Gateways

G

GartnerGroup
    e-mail cost study, 35-36
    LAN cost study, 36
    OIS migration cost study, 36
Gateways, 6, 8-9, 23-24, 40-47, 163-171, 256-265
    common features and services, 45-49
        access to utilities, 49
        administration, configuration, and system
            monitoring, 48
        centralized directory facility, 45
        directory synchronization, 45
        document conversion, 49
        security and encryption, 48
    defined, 6
    directory and transport backbone
        components, 260
    distributed multisystem gateways, 44, 46-47
    functionality tests, 9-10, 163-171
        address mapping, 164
        alternate recipient allowed, 166
        autoforward indication, 166-167
        blind copy recipient handling, 167
        body part encryption indication, 167
        character set mapping, 164
        content information, 166
        content type, 167
        conversion prohibition, 167
        conversion prohibition in case of loss of
            information, 167-168
        converted indication, 168

        cross-referencing indication, 168
        delivery notification, 168
        delivery via fax, telex, post, 169
        DL expansion history indication, 169-170
        DL expansion prohibited, 170
        express mail service (EMS), 170
        limitations on number of addresses, 165
        message unique ID, 166
        multipart body, 170
        no nondelivery notification, 169
        nondelivery notification, 169
        overview of, 9-10, 163-164
        priority systems, 166
        problem resolution, 170-171
        read receipt, 168-169
        return of content with nondelivery
            notification, 169
        subject field contents, 165
        subject field length, 165
        time stamps, 165
        trace strings, 166
    multigateway distributed switches, 259-260
    multigateway switches, 257-258
    multisystem gateways, 43-44
    overview of, 8-9, 23-24, 40-41, 256-257
    product comparison chart, 261-265
    product-independent message store
        products, 260-261
    single gateways, 41-43
    stand-alone gateways, 257, 258
    supported in
        ALL-IN-1 software, 102-103
        BeyondMail, 73-74
        cc:Mail, 140
        DaVinci eMail, 87
        Emc^2/TAO, 108
        Lotus Notes, 148
        Microsoft Mail, 156
        OfficeVision/400, 129-130
        OfficeVision/VM, 126
        Open DeskManager, 118
        QuickMail, 82
        RFP/RFI example, 308
    to value added networks (VANs), 55-56
        Internet gateways, 56-57
        overview of, 55-56
        proprietary gateways, 56
        X.400 gateways, 56
    See also Interoperability; Value added
        networks (VANs)
Global Message Handling Service (Global MHS).
    See NetWare Global Message Handling Service
Graphics, document conversion and, 273
Group Conferencing feature, ALL-IN-1, 91

H

Headings. *See* Message headings
Hewlett Packard. *See* Open DeskManager software
Hubs, in multisystem gateways, 6-7, 43-45

I

IBM. *See* Document Interchange Architecture
   (DIA); OfficeVision/400 software;
   OfficeVision/VM software; SNA Distribution
   Services (SNADS)
IDs, testing message unique ID, 166
IETF (Internet Engineering Task Force), 251,
   252-253
Impact, of e-mail, 32
Information discrepancies, in inverse mapping, 160
Infrastructure security, 276, 278
Installation specifications, in RFP/RFI example, 311
Interconnecting e-mail systems. *See* Interoperability
Interfaces. *See* Message user agents
International Federation of Information Processing
   (IFIP), 251
International Organization for Standardization
   (ISO), 251-252
      See also Open Systems Interconnection (OSI)
         protocol stack
Internet addressing, across VANs, 59
Internet Engineering Task Force (IETF), 251,
   252-253
Internet gateways, to value added networks
   (VANs), 56-57
Internet message mapping, 160-163
      body part mapping, 162-163
      heading and envelope mappings, 160-162
Interoperability, 7-12, 158-171
      address mapping between systems, 159-160
      gateway functionality tests, 9-10, 163-171
         address mapping, 164
         alternate recipient allowed, 166
         autoforward indication, 166-167
         blind copy recipient handling, 167
         body part encryption indication, 167
         character set mapping, 164
         content information, 166
         content type, 167
         conversion prohibition, 167
         conversion prohibition in case of loss of
            information, 167-168
         converted indication, 168
         cross-referencing indication, 168
         delivery notification, 168
         delivery via fax, telex, post, 169
         DL expansion history indication, 169-170
         DL expansion prohibited, 170
         express mail service (EMS), 170

         limitations on number of addresses, 165
         message unique ID/trace strings, 166
         multipart body, 170
         no nondelivery notification, 169
         nondelivery notification, 169
         overview of, 9-10, 163-164
         priority systems, 166
         problem resolution, 170-171
         read receipt, 168-169
         return of content with nondelivery
            notification, 169
         subject field contents, 165
         subject field length, 165
         time stamps, 165
      nonsymmetric information mapping, 159
      overview of, 7-8, 158-159
      specifications in RFP/RFI example, 307
      X.400 and Internet message mapping,
         160-163
         body part mapping, 162-163
         heading and envelope mappings, 160-162
      See also E-mail systems; Gateways
Interpersonal Messaging (IPM), 220-221, 232-240
      message store (MS), 221, 240-241
      message structure, 220, 232-240
         Abstract Syntax Notation (ASN.1)
            encoding, 233-236, 239-240
         body, 238-239
         envelopes, 237
         headings, 237-238
         overview of, 232
         versions of, 220
      message transport agents (MTAs), 20-21, 222
      message user agent (MUA), 220-221
      overview of, 220-221
      See also X.400 standard
Inverse mapping, information discrepancies in, 160
ISO. *See* International Organization for
   Standardization; Open Systems Interconnection
   (OSI) protocol stack

L

Limitations on number of addresses, 165
List CMC function, 287, 289
Local area network (LAN) cost studies, 33-36
      Ferris Network LAN e-mail cost study, 33-35
      Forrester LAN cost study, 33
      GartnerGroup LAN cost study, 36
      GartnerGroup OIS migration cost study, 36
Looking up names, with CMC APIs, 287
Lotus Notes software, 141-148
      addressing structure, 144-145
      contact information, 141
      directory structure, 143-144
      gateways supported, 148

mail delivery, 145-146
message store, 148
message structure, 146-148
message transport system, 145-146
message user agent, 142-143
Name & Address Book, 143-144
network specifications, 142
operating system specifications, 142
overview of, 141-142
platform specifications, 142
See also cc:Mail software

M

Macintosh computers
cc:Mail for, 133
QuickMail for, 76
Mail delivery. *See* Message delivery
Mainframe-based e-mail systems, 36
Management
ALL-IN-1 System Management utilities, 91
ALL-IN-1 Time Management feature, 90
cc:Mail time management feature, 133
OfficeVision/VM distribution manager, 124
X.500 Administrative Directory
Management Domain (ADDMD), 180
X.500 Private Directory Management
Domain (PRDMD), 180
See also Network management
Management information bases (MIBs), 252-253
MAPI. *See* Messaging Application Program
Interface (MAPI) API
Mapping
address mapping, 159-160, 164
gateway functionality test, 164
information discrepancies in inverse
mapping, 160
character set mapping, 164
information between MIME and X.400 body
parts, 158
nonsymmetric information mapping, 159
X.400 and Internet message mapping,
160-163
body part mapping, 162-163
heading and envelope mappings, 160-162
See also Interoperability
Message attachments
in ALL-IN-1, 92
in Microsoft Mail, 155
in Open DeskManager, 111
overview of, 26
See also Message structures
Message body
body part encryption indication service, 167
in cc:Mail, 140
defined, 26-28

in DIA, 191-192
extended body parts (EBPs), 162
in Interpersonal Messaging (IPM), 238-239
in NetWare Global Message Handling
Service (MHS), 200
in OfficeVision/VM, 125-126
testing for multipart body, 170
See also Message structures
Message definition call, CMC, 288
Message delivery
via fax, telex, and postal service, 169
forms supported in RFP/RFI example,
309-310
in Lotus Notes, 145-146
notification services, 93, 168-169
in X.400 standard, 231, 233
Message envelopes
in BeyondMail, 70-72
in Interpersonal Messaging (IPM), 237
in NetWare Global Message Handling
Service (Global MHS), 196-198
in OfficeVision/VM, 125
overview of, 26-27, 29-30
in SNADS and DIA, 190
See also Message structures
Message functions, ALL-IN-1, 92
Message Handling Service (MHS). *See* NetWare
Global Message Handling Service
Message headings
in cc:Mail, 139
defined, 26-27
in Interpersonal Messaging (IPM), 237-238
in NetWare Global Message Handling
Service (Global MHS), 198-200
in OfficeVision/VM, 125
in QuickMail, 81
in SNADS and DIA, 190-191
See also Message structures
Message Router OpenVMS gateway, ALL-IN-1, 90,
95-97
Message security services, 277-278
See also Security
Message store (MS)
ALL-IN-1, 102
BeyondMail, 73
cc:Mail, 140
DaVinci eMail, 86
directory information sharing and, 18-19
Interpersonal Messaging (IPM), 221, 240-241
Lotus Notes, 148
Microsoft Mail, 155
OfficeVision/400, 129
OfficeVision/VM, 126
Open DeskManager, 118
overview of, 21-22
product-independent message store
gateway products, 260-261

QuickMail, 82
Message structures, 5-6, 25-30
    ALL-IN-1, 97-102
    attachments, 26
    BeyondMail, 69-73
    cc:Mail, 139-140
    DaVinci eMail, 86
    Document Interchange Architecture (DIA),
        189-192
    Emc^2/TAO, 108
    Interpersonal Messaging (IPM), 220, 232-240
    Lotus Notes, 146-148
    mapping information between MIME and
        X.400 body parts, 158
    Microsoft Mail, 154-155
    NetWare Global Message Handling Service
        (Global MHS), 195-200
    notification messages, 28
    OfficeVision/400, 129
    OfficeVision/VM, 124-126
    Open DeskManager, 117-118
    overview of, 5-6
    QuickMail, 81-82
    RFC 822 message structure, 211, 212, 215
    Simple Message Transfer Protocol (SMTP),
        211-217
    SNA Distribution Services (SNADS), 189-192
    types of messages, 25
    VIM APIs and, 283
    X.400 probe messages, 25
    See also Interoperability; Message
        attachments; Message body; Message
        envelopes; Message headings;
        Multipurpose Internet Mail Extension
        (MIME)
Message switches, 6-7, 43-44
Message transport agents (MTAs)
    Interpersonal Messaging (IPM) and, 20-21,
        222
    overview of, 14-16, 20-21
Message transport systems (MTSs)
    ALL-IN-1, 95-97
    BeyondMail, 68-69, 70
    cc:Mail, 138-139
    DaVinci eMail, 86
    defined, 6, 20
    Emc^2/TAO, 108
    Lotus Notes, 145-146
    Microsoft Mail, 153-154
    NetWare Global Message Handling Service
        (Global MHS), 200-201
    OfficeVision/400, 129
    OfficeVision/VM, 123-124
    Open DeskManager, 116-117
    QuickMail, 78-80
    Simple Message Transfer Protocol (SMTP),
        208-211

    SNADS and DIA, 189
    X.400 standard, 228-232
Message unique IDs, 166
Message user agents (MUAs)
    ALL-IN-1, 94
    BeyondMail, 66
    cc:Mail, 134
    DaVinci eMail, 85
    Emc^2/TAO, 106
    Interpersonal Messaging (IPM), 220-221
    Lotus Notes, 142-143
    Microsoft Mail, 151
    OfficeVision/400, 128
    OfficeVision/VM, 122
    Open DeskManager, 111
    overview of, 14-16, 22-23
    QuickMail, 76
Messages
    mapping information between MIME and
        X.400 body parts, 158
    notification messages, 28
    in OfficeVision/400, 127
    in OfficeVision/VM, 120-121
    X.400 probe messages, 25
Messaging Application Program Interface (MAPI)
    API, 290-292
    MAPIAddress function, 291
    MapiFileDesc definition, 291-292
    MapiMessage structure definition, 291, 292
    MapiRecipDesc definition, 291
    MAPISendMail function, 291
    overview of, 290-291
    See also Application program interfaces
Messaging calls, CMC, 287-289
MHS (Message Handling Service). *See* NetWare
    Global Message Handling Service
MIBs (management information bases), 252-253
Microsoft Mail software, 148-156
    addressing structure, 152, 155
    attachments, 155
    bulletin board support, 149
    calendaring feature, 149
    contact information, 149
    directory structure, 151-152
    directory updating, 152
    document conversion feature, 150
    EXTERNAL program, 153-154
    file cabinet feature, 150
    forms feature, 149-150
    gateways supported, 156
    message store, 155
    message structure, 154-155
    message transport system, 153-154
    multimedia mail feature, 150
    network specifications, 150
    operating system specifications, 150
    overview of, 148-150

platform specifications, 150
remote user access feature, 150
Schedule+ and, 149
user agent, 151
WinRules feature, 150
workgroup templates feature, 150
Microsoft Windows
cc:Mail for, 132
features in BeyondMail, 65
See also Messaging Application Program
Interface (MAPI) API
Migration cost study, GartnerGroup, 36
MIME. *See* Multipurpose Internet Mail Extension
Mobile mail support, cc:Mail, 133
Monitoring tools, gateways and, 48
MS-DOS. *See* DOS
MTAs (message transport agents), 14-16, 20-21, 222
MTSs. *See* Message transport systems
MUAs. *See* Message user agents
Multicasting, in X.500 directory standard, 184, 185
Multigateway distributed switches, 259-260
Multigateway switches, 257-258
Multilingual support. *See* Foreign language
support
Multimedia mail feature, Microsoft Mail, 150
Multipart body, testing for, 170
Multiple columns, document conversion and, 272
Multipurpose Internet Mail Extension (MIME)
message structure, 158, 211-217
body parts, 212-215
boundary strings, 212-213
data encoding techniques, 215-217
mapping information between X.400 body
parts and, 158
overview of, 211-212
See also Message structures; Simple Message
Transfer Protocol
Multisystem gateways, 43-44, 46-47

N

Name & Address Book, Lotus Notes, 143-144
Names, looking up with CMC APIs, 287
NetWare Global Message Handling Service
(Global MHS), 21, 194-203
addressing in, 194-195
body, 200
directory structure, 201-203
envelopes, 196-198
headings, 198-200
message structure, 195-200
message transport system, 200-201
overview of, 21, 194
Simple Message Transfer Protocol (SMTP)
and, 194-195
NetWare Loadable Modules (NLMs), 21

Network layer, OSI protocol stack, 247
Network management, 10-12, 250-254
configuration management, 252
fault management, 252
frameworks, 251-252
management information bases (MIBs),
252-253
overview of, 10-12
performance management, 252
security management, 252
standards, 250-251
See also Management
Network Scheduler 3 software, 75
Network Services (NS) protocol, Open
DeskManager, 116-117
Network specifications
ALL-IN-1, 94
BeyondMail, 65-66
cc:Mail, 134
DaVinci eMail, 85
Emc^2/TAO, 105
Lotus Notes, 142
Microsoft Mail, 150
OfficeVision/400, 128
OfficeVision/VM, 122
Open DeskManager, 111
QuickMail, 76
Networks. *See* Local area network (LAN) cost
studies; Value added networks
Nickname support, in OfficeVision/VM, 121
NLMs (NetWare Loadable Modules), 21
No nondelivery notification services, 169
Nondelivery notification services, 169
Nonsymmetric information mapping, 159
Notes
document conversion and, 272
in OfficeVision/400, 127
in OfficeVision/VM, 120-121
See also Documents; Messages
Notification messages
in ALL-IN-1, 93
delivery notification services, 168
no nondelivery notification service, 169
nondelivery notification service, 169
overview of, 28
return of content with nondelivery
notification service, 169
Novell. *See* DaVinci eMail; NetWare Global
Message Handling Service
NS (Network Services) protocol, Open
DeskManager, 116-117

O

OfficeVision/400 software, 126-131
addressing structure, 129

bulletin board support, 128
calendaring feature, 127-128
contact information, 126
directory structure, 129-130
gateways supported, 129-130
message store, 129
message structure, 129
message transport system, 129
message user agent, 128
messages, notes, and documents, 127
network specifications, 128
operating system specifications, 128
overview of, 127-128
platform specifications, 128
OfficeVision/VM software, 118-127
addressing structure, 123
bulletin board support, 121
calendaring feature, 120
CallUp directory, 122
contact information, 119
directory record format, 123
directory structure, 122-123
directory synchronization, 122
DisplayWrite/370 document conversion, 121-122
distribution manager, 124
DOS Office Direct Connect feature, 121
file cabinet feature, 121
file format, 123
gateways supported, 126
message body, 125-126
message envelopes, 125
message headings, 125
message store, 126
message structure, 124-126
message transport system, 123-124
message user agent, 122
messages, notes, and documents, 120-121
network specifications, 122
nickname support, 121
operating system specifications, 122
overview of, 118-120
platform specifications, 122
record field types, 122-123
OIS migration cost study, GartnerGroup, 36
OLE support, in BeyondMail, 65
Online services, versus value added networks (VANs), 55, 57
Online training feature, ALL-IN-1, 91
Open DeskManager software, 108-118
addressing structure, 115-116
attachments, 111
contact information, 109
directory structure, 111-115
Distributed Services (DS) protocol, 116
file cabinet feature, 111
forms routing feature, 111

gateways supported, 118
message store, 118
message structure, 117-118
message transport system, 116-117
message user agent, 111
Network Services (NS) protocol, 116-117
network specifications, 111
new features, 110
operating system specifications, 111
overview of, 108-110
platform specifications, 111
printing features, 110
Open Systems Interconnection (OSI) protocol stack, 244-248
Application layer, 245-246
Data-link layer, 247-248
Network layer, 247
overview of, 244-245
Physical layer, 248
Presentation layer, 246
Session layer, 246
Transport layer, 246-247
See also Security
Operating system specifications
ALL-IN-1, 93
BeyondMail, 65-66
cc:Mail, 133
DaVinci eMail, 85
Emc^2/TAO, 105
Lotus Notes, 142
Microsoft Mail, 150
OfficeVision/400, 128
OfficeVision/VM, 122
Open DeskManager, 111
QuickMail, 76
Organizational impact, of e-mail, 32
OS/2 Workplace Shell, cc:Mail for, 132-133
OSI. *See* Open Systems Interconnection (OSI) protocol stack
Outside/In Views, BeyondMail, 64

P

P2 protocol, 220
See also Interpersonal Messaging (IPM)
Page layout, document conversion and, 271
Paragraph information, document conversion and, 272
Paragraph styles, document conversion and, 272
PC services, in RFP/RFI example, 309
Performance management, 252
Personal address books, 24, 143-144
Physical layer, OSI protocol stack, 248
Platform specifications
ALL-IN-1, 93
BeyondMail, 65-66

cc:Mail, 133
DaVinci eMail, 85
Emc^2/TAO, 105
Lotus Notes, 142
Microsoft Mail, 150
OfficeVision/400, 128
OfficeVision/VM, 122
Open DeskManager, 111
QuickMail, 76
Post office (PO). *See* Message store
Postal service, testing delivery via, 169
Presentation layer, OSI protocol stack, 246
Printing features
    ALL-IN-1, 90, 91
    Open DeskManager, 110
Priority systems, 166
Private Directory Management Domain (PRDMD),
    X.500, 180
Private encryption keys, 276-277
Probe messages, X.400, 25
Problem resolution, after gateway functionality
    tests, 170-171
Product-independent message store gateway
    products, 260-261
Proprietary addressing, across VANs, 59
Proprietary gateways, to value added networks
    (VANs), 56
Protocols
    defined, 5
    Directory Access Protocol (DAP), 179
    Directory Information Shadowing Protocol
      (DISP), 179
    Directory System Protocol (DSP), 179
    Distributed Services (DS) protocol, 116
    Extended Simple Message Transfer Protocol
      (SMTP), 210-211
    Network Services (NS) protocol, 116-117
    P2 protocol, 220
    See also Open Systems Interconnection (OSI)
      protocol stack; Simple Message Transfer
      Protocol (SMTP)
Public encryption keys, 48, 276-277

Q

Queries, directory system agents and, 183-184, 185
QuickMail software, 74-82
    addressing structure, 77-78
    contact information, 74
    directory structure, 76-77
    gateways supported, 82
    heading fields, 81
    for the Macintosh, 76
    message store, 82
    message structure, 81-82
    message transport system, 78-80

    message user agent, 76
    Network Scheduler 3 software and, 75
    network specifications, 76
    operating system specifications, 76
    overview of, 75-76
    platform specifications, 76
    QM Forms software, 75
    QM Remote software, 75
    volume directory structure, 78-80

R

Read CMC function, 287, 289
Read receipt services, 168-169
Receipt notification feature, ALL-IN-1, 93
Record field types, OfficeVision/VM, 122-123
Recovery specifications, in RFP/RFI example, 308
Referral, in X.500 directory standard, 184, 185
Remote user access feature
    Microsoft Mail, 150
    QuickMail, 75
Request For Proposal/Request For Information
    (RFP/RFI) example, 306-315
    administrative specifications, 310-311
    backup and recovery specifications, 308
    costs, 312-315
    customer support specifications, 311-312
    delivery forms supported, 309-310
    gateway support, 308
    installation specifications, 311
    interoperability specifications, 307
    PC services, 309
    security specifications, 308-309
    standards support, 308
    technical specifications, 306-309
Return of content with nondelivery notification
    services, 169
RFC 822 message structure, 211, 212, 215

S

Schedule+ software, Microsoft Mail and, 149
Scripting languages, BeyondMail, 65
Section information, document conversion and, 271
Security, 276-279
    access control, 276-277
    authenticated and authorized users, 276
    in BeyondMail, 64
    e-mail components and, 276
    encryption, 48, 167, 276-277
    end-to-end security, 276, 278
    gateways and, 48
    infrastructure security, 276, 278
    management of, 252
    message security services, 277-278
    specifications in RFP/RFI example, 308-309

Semicolon (;), in X.400 addressing, 228
Send CMC function, 287, 288-289
Services
  defined, 7, 164
  See also Gateways, functionality tests
Session layer, OSI protocol stack, 246
Set Mail User option, ALL-IN-1, 89
Shared directories, 16-19
  global distributed directories, 18
  global nondistributed directories, 18
  maintaining global directory information,
    18-19
  overview of, 16-17
  See also Directories
Shared filing feature, ALL-IN-1, 89
Simple Message Interface (SMI) APIs, VIM, 283-284
Simple Message Transfer Protocol (SMTP), 206-217
  address directory, 206, 217
  converting binary data to ASCII data, 211,
    215-217
  Domain Name Service (DNS) addressing
    structure, 206-207
  Extended SMTP, 210-211
  message structures, 211-217
  message transport system, 208-211
  Multipurpose Internet Mail Extension
    (MIME) message structure, 158, 211-217
    body parts, 212-215
    boundary strings, 212-213
    data encoding techniques, 215-217
    mapping between X.400 body parts and,
      158
    overview of, 211-212
  NetWare Global Message Handling Service
    and, 194-195
  overview of, 20, 206
  RFC 822 message structure, 211, 212, 215
  top level domains (TLDs), 206-207
Single gateways, 41-43
Slash (/), in X.400 addressing, 228
SMF (Standard Message Format), 194
  See also NetWare Global Message Handling
    Service
SMI (Simple Message Interface) APIs, VIM, 283-284
SNA Distribution Services (SNADS), 21, 188-192
  address directory, 192
  addressing in, 188
  message structure, 189-192
  message transport system, 189
  overview of, 21, 188
  See also Document Interchange Architecture
    (DIA)
Stand-alone gateways, 257, 258
Standard Message Format (SMF), 194
  See also NetWare Global Message Handling
    Service

Standards
  ALL-IN-1 support for, 93
  for network management, 250-251
  support for in RFP/RFI example, 308
  See also Protocols; X.400 standard; X.500
    directory standard
Subject field
  testing contents, 165
  testing length, 165
Switches
  e-mail switches, 6-7, 43-44
  multigateway distributed switches, 259-260
  multigateway switches, 257-258
System Management utilities, ALL-IN-1, 91
System monitoring, with gateways, 48
Systems Network Architecture Distribution
  Services (SNADS). *See* SNA Distribution
  Services

T

Tables, document conversion and, 272
TeamLinks support, ALL-IN-1, 91-92
Technical specifications, in RFP/RFI example,
  306-309
Telecommunications carriers. *See* Value added
  networks (VANs)
Telex, testing delivery via, 169
Testing. *See* Gateways, functionality tests
Time management features
  ALL-IN-1, 90
  cc:Mail, 133
  See also Calendaring features
Time stamps, 165
Top level domains (TLDs), 206-207
Trace strings, 166
Training feature, ALL-IN-1, 91
Transport backbone and directory gateway
  components, 260
Transport layer, OSI protocol stack, 246-247
Transports. *See* Message transport agents (MTAs);
  Message transport systems (MTSs)

U

UNIX, cc:Mail for, 133
User agents. *See* Message user agents
User profile feature, ALL-IN-1, 90
UUENCODE/UUDECODE utility, 211, 215

V

Value added networks (VANs), 7, 54-60
  addressing across VANs, 58-59
    Internet addressing, 59

proprietary addressing, 59
X.400 addressing, 59, 226-228
criteria for selecting, 60
defined, 7, 54
Electronic Data Interchange (EDI) and
e-mail services, 54, 57
gateways to, 55-56
Internet gateways, 56-57
overview of, 55-56
proprietary gateways, 56
X.400 gateways, 56
versus online services, 55, 57
overview of, 54
vendors, 58
See also Gateways; Request For
Proposal/Request For Information
(RFP/RFI) example
Vendor Independent Messaging (VIM) API,
282-286
address format, 286
character sets, 283
data attribute encoding, 283
message creation process, 284-286
message structure, 283
overview of, 282-283
Simple Message Interface (SMI) APIs,
283-284
VIMCreateMessage function, 284
VIMSetMessageHeader function, 284-285
VIMSetMessageItem function, 285-286
VIMSetMessageRecipient function, 285
See also Application program interfaces
Vendors, of value added networks (VANs), 58
Volume directory structure, QuickMail, 78-80

W

Windows. *See* Microsoft Windows
WinRules feature, Microsoft Mail, 150
Workgroup templates feature, Microsoft Mail, 150

X

X.200 standard. *See* Open Systems Interconnection
(OSI) protocol stack

X.400 standard, 220-240
address directory, 241
addressing, 59, 222-228
Form 1, Variant 1 addressing, 223-226
between systems, 59, 226-228
types of, 222-223
e-mail systems and, 14-16
gateways to value added networks (VANs),
56
mapping information between MIME and
X.400 body parts, 158
message mapping, 160-163
body part mapping, 162-163
heading and envelope mappings, 160-162
message transport agent (MTA), 20-21
message transport system, 228-232
message delivery, 231, 233
message submission, 228-229, 230
message transfer, 229-231
Probe messages, 25
versus X.500 directory standard, 178-179
See also Interpersonal Messaging (IPM);
Security
X.500 directory standard, 17, 176-185
Administrative Directory Management
Domain (ADDMD), 180
chaining in, 183-184
components and protocols, 179-180
Directory Access Protocol (DAP), 179
Directory Information Shadowing Protocol
(DISP), 179
directory information tree (DIT), 181
Directory System Protocol (DSP), 179
directory user agent (DUA) and directory
system agent (DSA) communications,
183-184, 185
distinguished names, 182
e-mail systems and, 15
implementation of, 176-179
multicasting in, 184, 185
overview of, 17, 176-179
Private Directory Management Domain
(PRDMD), 180
referral in, 184, 185
versus X.400 standard, 178-179

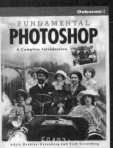

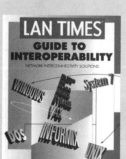

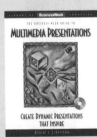

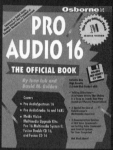

Yo Unix!

open COMPUTING

INNOVATIVE BOOKS

FROM OPEN COMPUTING AND

OSBORNE/MCGRAW-HILL

open COMPUTING

Guide to THE BEST FREE UNIX UTILITIES

Includes One CD

James Keough & Remon Lapid

OPEN COMPUTING'S
GUIDE TO THE BEST
FREE UNIX UTILITIES
BY JIM KEOUGH AND
REMON LAPID
INCLUDES
ONE CD-ROM
$34.95 U.S.A.
ISBN: 0-07-882046-4
AVAILABLE NOW

OPEN COMPUTING'S
BEST UNIX TIPS
EVER
BY KENNETH H. ROSEN,
RICHARD P. ROSINSKI,
AND DOUGLAS A. HOST
$29.95 U.S.A.
ISBN: 0-07-881924-5
AVAILABLE NOW

OPEN COMPUTING'S
UNIX UNBOUND
BY HARLEY HAHN
$27.95 U.S.A.
ISBN: 0-07-882050-2

OPEN COMPUTING'S
STANDARD UNIX
API FUNCTIONS
BY GARRETT LONG
$39.95 U.S.A.
ISBN: 0-07-882051-0

BC640SL